Neurology and Modernity

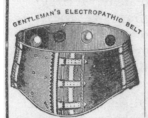
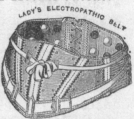

Advertisement, John Strange Winter, *My Poor Dick* (London: VF White & Co, 1892), 111

Neurology and Modernity

A Cultural History of Nervous Systems, 1800–1950

Edited by

Laura Salisbury
RCUK Fellow in Science, Technology and Culture, Birkbeck College, London

Andrew Shail
Lecturer in Film, Newcastle University

First published 2010 by
PALGRAVE MACMILLAN

Palgrave Macmillan in the UK is an imprint of Macmillan Publishers Limited, registered in England, company number 785998, of Houndmills, Basingstoke, Hampshire RG21 6XS.

Palgrave Macmillan in the US is a division of St Martin's Press LLC, 175 Fifth Avenue, New York, NY 10010.

Palgrave Macmillan is the global academic imprint of the above companies and has companies and representatives throughout the world.

Palgrave® and Macmillan® are registered trademarks in the United States, the United Kingdom, Europe and other countries.

ISBN-13: 978–0–230–23313–3 hardback

This book is printed on paper suitable for recycling and made from fully managed and sustained forest sources. Logging, pulping and manufacturing processes are expected to conform to the environmental regulations of the country of origin.

A catalogue record for this book is available from the British Library.

A catalog record for this book is available from the Library of Congress.

10   9   8   7   6   5   4   3   2   1
19 18 17 16 15 14 13 12 11 10

Printed and bound in Great Britain by
CPI Antony Rowe, Chippenham and Eastbourne

Contents

List of Figures

Acknowledgements

Laura Salisbury would like to thank her colleagues Roger Luckhurst and Gill Partington for their invaluable advice and support at various stages of this project. Hilary Fraser also offered much encouragement, just when it was most needed. Isabel Davis offered publishing advice, friendship and delicious meals, all of which contributed to the completion of this book. Finally, I would like to thank Adam Turnock for his unstinting care.

Andrew Shail would like to thank all at Special Collections at Newcastle University's Robinson Library for help above and beyond, and Stacy Gillis and Bean Shail-Gillis for being delicious.

Laura and Andrew would like to thank the contributors to this volume. All the scholars involved displayed remarkable patience and attention to detail, and some of them offered much practical support. We would like to thank them all for their belief in the project.

Copyright Acknowledgements

Notes on Contributors

Michael K. House is a graduate student at Princeton University and is beginning an Andrew W. Mellon Postdoctoral Fellowship at the University of Toronto's Jackman Humanities Institute (2009–11). His dissertation, *Grounding Fictions: Systemic Skepticism and Critical Doubt 1792–1807*, looks at the common engagement with scepticism in Early German Romanticism and post-Kantian philosophy. He is currently working on two projects. The first examines the shift of the term *Schwärmerei* (enthusiasm or fanaticism) from religious to political discourse around 1800. The second addresses the formation of empirical psychology in Moritz's *Magazin zur Erfahrungsseelenkunde* in the context of emerging scientific models for the study of the human.

Hisao Ishizuka is Associate Professor in the Department of English at Senshu University, Japan. He has published articles in *History of Science* and *Literature and Medicine*, and most recently contributed to *Liberating Medicine, 1720–1835* (2009). He is co-editor of *Shintai Ibunkaron* [*Body, Medicine and Culture*] (2002). He is currently writing on fibre theory in Enlightenment medicine, William Blake and medical sciences and the medico-cultural experience of dyspepsia in nineteenth-century Britain.

Melissa M. Littlefield is an assistant professor in the Departments of English and Kinesiology & Community Health at the University of Illinois, Urbana-Champaign. She is currently a faculty fellow at the Illinois Program for Research in the Humanities. Her work has appeared in the journal *Science, Technology and Human Values*. She is the author of a monograph on the cultural history of deception detection, and the co-editor, with Susan Squier, of a special issue of *Feminist Theory* on 'Feminist Theory and/of Science'. Her current projects address metadisciplinarity in the forensic sciences and the neuroscience of science fiction's future justice systems.

Jessica Meyer completed her PhD on 'The First World War and Narratives of Heroic and Domestic Masculinity in Britain, 1915–1937' at the University of Cambridge in 2005. She holds an MPhil in European Studies from the University of Cambridge and a BA from Yale University. Her publications include *Men of War: Masculinity and the First World War in*

Britain (2009), articles on shell shock, martial masculinity and the experiences of wives of disabled ex-servicemen, and two edited collections, *British Popular Culture and the First World War* (2008) and *Masculinity and the Other: Historical Perspectives* (2009).

Vike Martina Plock is a lecturer in English literature at Northumbria University. She is the author of *Joyce, Medicine, and Modernity* (2009) and has written widely on James Joyce, modernism and science. She is currently guest-editing a special issue of *James Joyce Quarterly* on Joyce and physiology and is also working on a new research project on women's writing, fashion and literary modernity.

George Rousseau is a fellow of the Royal Historical Society and the recipient of many honorary degrees *honoris causa*. He has been a professor at UCLA, Regius professor of English literature at King's College Aberdeen and Co-Director of the Centre for the History of Childhood at Oxford University. He now lives and writes in Oxfordshire. Among his books is a trilogy about Enlightenment culture – *Enlightenment Borders*, *Enlightenment Crossings* and *Perilous Enlightenment* (all 1991) – as well as *Nervous Acts: Essays on Literature, Culture and Sensibility* (2004) and *Children and Sexuality: The Greeks to the Great War* (2007). His biography of Sir John Hill entitled *Notorious* will soon be published.

Laura Salisbury is a lecturer and Research Councils UK research fellow in Science, Technology and Culture at Birkbeck, University of London. She has published on Samuel Beckett, ethics and comedy and on literature and philosophy. Her forthcoming work includes a monograph on Beckett, a volume on late modernism and a co-edited collection on the work of Friedrich Kittler. Her major research project is on the relationship between neurological conceptions of language, modernism and modernity.

Aura Satz is a fellow at the London Consortium. She completed her PhD at the Slade School of Fine Art, where she subsequently held a Henry Moore post-doctoral fellowship (2002–4). She is co-editor (with Jonathan Wood) of *Articulate Objects: Voicing and Listening to Sculpture and Performance* (2009), and has published essays in books and journals on puppets, tableaux vivants, iconoclasm, automata, phantom limbs and spiritualism. As an artist she has shown and performed her work in the United Kingdom and internationally, including FACT Liverpool, Site Gallery, Whitechapel Gallery, The Photographers' Gallery, the Victoria and Albert Museum, Tate Britain and Beaconsfield Gallery. Her projects can be seen on www.iamanagram.com

Andrew Shail is Lecturer in Film at Newcastle University. He is co-author of a BFI Film Classic on *Back to the Future* (2010), editor of *Reading the Cinematograph* (2010) and a special issue of *Early Popular Visual Culture* on intermediality in early cinema (2010), and co-editor of *Menstruation: A Cultural History* (2005). He is the author of forthcoming monographs on the place of cinema in the emergence of literary modernism and on cinema in the life of HG Wells, and of numerous articles on the beginnings of British film culture. He is currently working on the origins of film celebrity in Britain, and a history of menstruation since 1750.

Michael Angelo Tata is Editor-in-Chief of advertising and fine-arts consulting group iPublishing (New York/Los Angeles), and serves on the editorial boards of journals *Kritikos* (United States) and *Nebula* (Australia), as well as *Doodlescope*, an online project for theorizing graffiti, doodling, automatic writing and other marginal practices of marking (University of Macao, China). His first book, *Andy Warhol: Sublime Superficiality*, is forthcoming in 2009. He has contributed to numerous critical anthologies, including *Madonna's Drowned Worlds* (2004) and *The Globetrotting Shopaholic* (2008). His most recent poetry appears in *Gertrude* (2009, ed. Steven Rydman). He is currently working on critical pieces about the aesthetics of forgiveness in Oscar Wilde's *De Profundis*, Lacanian object relations and Jacques Derrida's concept of the Gift of Death, all for inclusion in various anthologies and journals.

Jane F. Thrailkill is Associate professor of English and comparative literature at the University of North Carolina, Chapel Hill. Her first book, *Affecting Fictions* (2007), draws on neuroscience to affirm the emotive and aesthetic dimensions of realist literature. She is at work on a monograph entitled *Talking Back: American Literature and the Subversive Child*, which examines how child consciousness is figured in poetry and narrative fiction. Her articles on the intersections of science, philosophy and literature have appeared in *English Literary History, American Literature, Studies in American Fiction*, and a number of collections.

Shelley Trower is AHRC research fellow at the University of Exeter. She is the editor of a special issue of *The Senses and Society* on 'Vibratory Movements' (2008), and has published several articles, including most recently in *Romanticism and Victorianism on the Net* (2009). She is currently writing a monograph on the relations between geology and ghost stories. She is also a committee member and publicity officer for the Oral History Society.

Jean Walton is Professor of English, Film/Media and Women's Studies at the University of Rhode Island. She is author of *Fair Sex, Savage Dreams: Race, Psychoanalysis, Sexual Difference* (2001), as well as articles in *Critical Inquiry*, *Discourse*, *differences*, *New Orleans Review*, *Contemporary Literature* and *College Literature*. She recently finished a novel set in Western Canada in the early 1970s titled *All Fine Motel*. This chapter is part of a longer study of peristaltic time and the body.

Introduction

Laura Salisbury and Andrew Shail

This book argues that to speak of neurology *and* modernity is to describe
a relationship of mutual constitution. As each of these chapters proposes, neurological conceptions of the self were primary components of
the ways in which 'modernity' – or the historical period usually seen to
stretch from 1850 to 1950 – conceptualized itself and its subjects. At the
same time, *Neurology and Modernity* explores the various ways in which
the historical characteristics of a broadly conceived cultural modernity
appear to have guided fundamentally the inquiries and discoveries of
the medical discipline of neurology. To begin to chart this complex and
knotty relationship between neurology and modernity is not simply to
suggest that the neurological self was a corollary of modernity, nor is it
to propose that neurology, as a discipline, was just the product of the
emergence of historical modernity; rather, it is to explore the ways in
which the two were symbiotically related, complexly co-generative.

In 'The Painter of Modern Life' (1859–60), one of the most significant descriptions of the relationship between modern experience and
aesthetics, Charles Baudelaire offered a by now familiar account of the
experience of modernity as exposure both to shocks and to an evanescence that defined a new sensation of the intensity of the present
moment. "[T]hat indefinable something we may be allowed to call
'modernity'", as Baudelaire puts it, is represented in experiential terms
according to the flickers and fluctuations of electrical stimulation; the
urban "lover of universal life moves into the crowd as though into an
enormous reservoir of electricity" (402, 400). But if experience is a category that mediates between the external world and the interiority of
the subject, then it is perhaps unsurprising that as without, so within.
The modern artist can capture the experience of modernity, "reflecting it at every moment in energies more vivid than life itself, always

1

inconstant and fleeting", precisely because s/he translates the sensations of the world into a body that vibrates and responds in sympathy with it, according to a model of nervous force (400). For Baudelaire, if "inspiration has some connection with congestion, ... every sublime thought is accompanied by a more or less vigorous nervous impulse that reverberates in the cortex"; as such, the artist, the "man of genius", necessarily has to have "strong nerves" in order to experience it sympathetically (398). This "man of genius" appears, at first, in contradistinction to "the child", whose nerves are weak, for "in the one, reason has assumed an important role; in the other, sensibility occupies almost the whole being" (398). But if artistic genius, in Baudelaire's terms, is the ability to recapture a childlike openness to sensation and impression whilst bringing "order into the sum of experience, involuntarily amassed" through the application of the "analytical mind", then the subjectivity of the modern artist is necessarily held in a tense oscillation between strong and weak nervous states. Baudelaire represents the artistic subject as functioning in a manner that is clearly analogous to the newly defined plasticity of the nervous system itself – a system that is both capable of receiving and giving or holding form – for he describes an artist who senses and is remoulded by the transitory shocks and impressions of the modern world, yet is able to fashion this evanescence into the permanence of the artwork. Caught between having weak and strong nerves, Baudelaire's artist needs to remain "perpetually in the spiritual condition of the convalescent", always in a somewhat nervous state, if he is to capture the essential quality of modernity (397). And Baudelaire's question of whether modernity vibrates in sympathy with a strong, vigorously nervous subject, or whether it finds its parallel in a nervous system defined by weakness and lassitude, traces out what became a central ambivalence in the modern era's attempts to conceptualize and describe its defining qualities.

The work in this volume follows Baudelaire's intuition of a sympathetic resonance between new configurations of a primarily urban, technological social world, and the shapes of a psychological and embodied subjectivity capable of translating external stimulation into internal processes, but it extends the reach of the argument significantly beyond the realm of the aesthetic. We argue that a broad range of cultural discourses that both arose from and described modernity can be thought of as being singularly neurological, determinedly nervous. One of the arguments that the chapters here make is that the ascription of the cause of somatic and psychological effects to the nerves was far more pervasive than within contemporary Western cultures. For example, in the popular fiction of the late-Victorian and Edwardian era, visceral states were described almost invariably in neurological terms, as the condition of the nerves was rendered fundamental to the experience and sometimes even the very perpetuation of the self. In an unpublished ending for *The Time Traveller's Story* (an earlier version of *The Time Machine*), whose serial publication in the *National Observer* was aborted in June 1894, H.G. Wells's narrator tells how the death of the Time Traveller, at the time known as 'the Philosophical Inventor', was ascribed by doctors to "stoppage of the heart, through some nervous lesion in the medulla" (TT-11). This lesion was 'nervous' not just in the sense of being 'related to a nerve', but nervous in the sense of being produced by nervous strain.

As other popular literary texts from the period show, the energy-supplying nerves became as vital to the proper functioning of the body as that which supplied the muscles. In Arnold Bennett's *The Old Wives' Tale* (1908), as Sophia reads a letter from her sister Constance, with whom she has had little contact during her adult life, she does not "betray physically that she was not reading an order for two rooms for a week. But the expenditure of nervous force necessary to self-control was terrific" (455). In a world where nervous strain was a permanent condition, to restore nervous energy was thus to restore health. In Grant Allen's 1896 short story 'An African Millionaire', Vandrift "recovers health and rejuvenates his nervous system by taking daily excursions along the coast to the Casion" (660) while staying in Nice, and in Wells's 1894 short story 'The Flowering of the Strange Orchid', Winter-Wedderburn is "a shy, lonely, rather ineffectual man, provided with just enough income to keep off the spur of necessity, and not enough nervous energy to make him seek any exacting employment" (9).

Because of the persistence of nervous strain, figured as a result of exposure to the modern world, to be 'nervous' was no longer to be sunk within a short-term anxiety; it was, instead, to be afflicted by a permanent tendency towards agitation. One could now be described as nervous without specific reference to an individual pathology. In William Le Queux's *The Count's Chauffeur* (1907), the chauffeur Ewart writes, when he tells of how he first suspected that Count Bindo and his companions were about to involve him in a crime, "I am not nervous by any means, yet I admit that at that moment I felt a decidedly uncomfortable feeling creeping over me" (16). Affect was also seen as the product of the expenditure of nervous energy. In *The Old Wives' Tale*, while Sophia is ruthlessly surviving the siege of Paris, "[t]he death of the faithful charwoman, when she heard of it, produced but little effect on Sophia, who was so overworked and so completely absorbed in her own affairs that she had no nervous energy to spare for sentimental regrets" (398). Nerves themselves also came to be generally understood to originate – rather than merely to communicate – a major portion of the body's sensations. In Wells's 'A Story of the Days to Come' (1899), when confronted by his violent workmates, "all the little nerves of Denton's being seemed leaping and dancing" (379). Irritation, of course, could be nervous, but, more surprisingly, so could violence, in that it was also a product of a transmission of nervous force. Sophia's tenant Chirac comes in one night, shutting the door with what is described as a "nervous violence" (412).

Thinking was similarly and consistently represented as a visceral event. In F. St Mars's 1911 short story 'Husband and Co.', the narrator recalls that on first encountering the beautiful wife of his friend Mr. Eest (whom Eest suspects of cheating on him), "like a flash there vibrated through my nerves a tingling consciousness of the power which this young lady wielded" (26). The body was perceived to be suffused with vibrating pathways, in a new figuration of the metaphorical vibration of plucked 'nerve strings'. In J.D. Beresford's *The Hampdenshire Wonder* (1911), replying negatively to inquiries about whether her genius child has ever spoken, Ellen Mary is seen by the narrator to be "twitching and vibrating. Her heavy, dark eyebrows jerked spasmodically, nervously" (16). And in John Buchan's *Prester John* (1910), David Crawfurd, hiding in the enemy army, faces discovery when each member of the army is called to swear a vow because his "nerves were already quivering" (113).

The nutritive processes of the body of modernity were generally understood to depend, in turn, on the nerves being properly fed. One 1916 advertisement described 'Phormoid' nerve tonic as "a natural nerve food; the nerves absorb it greedily, and as the nerves become stronger so the assimilative organs become more active, and food which formerly went to waste is used to nourish, strengthen, and build up the body" (9). Another advertisement from the same year insisted

> that insomnia, nervousness, lack of energy, neurasthenia, etc., can all be relieved and cured by the use of a food phosphate known among chemists as *bitro*-phosphate.... It feeds the nerves and makes them strong and steady. Vitality is raised to a high level, mental and physical lassitude are overcome; the brain acts quickly and clearly, and fatigue is less noticeable, even after a trying day (Bitro-Phosphate, 7, emphasis in original).

Debility was consequently figured as weakness of the nerves. A 'low fever' experienced by Rosamond, the protagonist of Mrs. Egerton Eastwick's 1896 short story 'One Season', is attributed by a doctor to "remaining in town during the summer heat, when already suffering from great nervous exhaustion" due to her husband's debts (286). In John Galsworthy's *Man of Property* (1906), Soames Forsyte "wished to God the house were finished, and they were in it, away from London. Town did not suit her [Irene]; her nerves were not strong enough" (174). Most of the fashionable disorders of the period were nervous. For the universally disapproving Lady Susan in H.H. Munro's 'A Matter of Sentiment' (1911), disapprobation was "what neuralgia and fancy needlework are to many other women" (138). Fashion and nervousness become aligned because the new experiences of modernity were commonly seen as making their most severe demands on the nerves. So in Wells's 1901 short story 'The New Accelerator', the inventor Gibberne, in concocting his nervous accelerator, can be described as "simply seeking an all-round nervous stimulant to bring languid people up to the stresses of these pushful days" (487).

If the experience of being modern was linked to a nervous system placed under increasing strain, it is no surprise that the evolutionary future was imagined according to modes of neurological adaptation. In *The War of the Worlds* (serialized April–December 1897) and *The First Men in the Moon* (serialized from November 1900 to August 1901), Wells, in envisioning the Martians and the Selenites as highly evolved races, imagined them as pure nervous system. Of the Martians, the narrator writes that "[t]he internal anatomy, I may remark here, as dissection has since shown, was … simple. The greater part of the structure was the brain, sending enormous nerves to the eyes, ear, and tactile tentacles" (125). Cavor describes the Grand Lunar, the ruler of the Selenites, as a "marvellous gigantic ganglion" (146), a vast nervous node unimpeded by a cranium. While humans possess bodies in which "digestive processes and their reaction upon the nervous system sap our strength and colour our minds", meaning that "[m]en go happy or miserable as they have healthy or unhealthy livers, or sound gastric glands", the narrator determines that "the Martians were lifted above all these organic fluctuations of mood and emotion" by their adapted, transcendentally neurological nature (*The War of the Worlds*, 125–6).

This pervasiveness of the neurological in allusions to the experience of modernity was, in one sense, a product of historical coincidence. Although the profession of neurology only became institutionalized in the United Kingdom in the period 1870–90,[1] the 'fathers' of modern neurology – as the first men to use the specialism 'neurologist' (in the 1830s) would later explain – had been writing in the 1770s and 1780s. It is nonetheless instructive that W.F. Bynum, a historian of neurology, cites Moritz Romberg's multi-volume *Manual of the Nervous Diseases of Man* (1840–6) as the *locus classicus* of modern neurology, specifically because, for the first time, it described 'nervous' diseases as consequences solely of affections of the nerves rather than as the result of pathological changes in their structure (92). The mid-nineteenth century, then, was the period when neurology became a distinct and, we will argue, a dominant mode for representing the more general vagaries of the embodied and mental life of the modern self. This collection thus explores the historical period 1800–1950 to describe neurology's ascendancy and to explore its relations to the era of modernity usually seen to straddle the nineteenth and twentieth centuries.

Two discourses

Modernity can be understood as a historical moment initially determined and then perpetuated by the effects of violent revolutions, secularization, the ascendancy of monopoly capitalism and bourgeois capitalist concepts of public and private, railway and then automotive transportation, the development of economies reliant on the industrialization of labour and the consequent rise of political structures for representing organized labour, and commercial dependence on long-distance instantaneous communication (via the telegraph). But just as significant to modernity as these material changes is the sense, persistently found in a broad range of discourses, of a common yet subjective experience of oneself *as* modern, as separated from the past to a degree not possible for one's ancestors. Modernity was persistently described and determined by a general sense of recent temporal rupture, accompanied by a consequent and self-conscious interest in new experiences, possibilities, technologies, dimensions, ways of sensing the world, and forms of representing it. Although it is possible and plausible to argue that modernity as a historical period had its beginnings far earlier than the second industrial revolution in Western Europe and North America, one of the things that makes the period with which *Neurology and Modernity* is concerned distinctive is its general tendency to conceptualize itself in terms of its radical difference from previous moments that were regarded more commonly, and by contrast, as having been experienced as part of a historical continuum.

In *We Have Never Been Modern*, the sociologist of science Bruno Latour argues that the 'modern Constitution' – that which produces society's sense of itself as modern – works obsessively at acts of 'purification', at categorizing the world according to a binary logic that separates the human from the non-human, culture from nature. To enter modernity, then, is allegedly to leave behind an impure world in which bodies, things and ideas are mixed up. As a consequence of this insistence upon a radical break, as Latour sees it,

> [t]he moderns have a peculiar propensity for understanding time that passes as if it were really abolishing the past behind it.... They do not feel that they are removed from the Middle Ages by a certain number of centuries, but that they are separated by Copernican revolutions, epistemological breaks, epistemic ruptures so radical that nothing of that past survives in them.... Since everything that passes is eliminated for ever, the moderns indeed sense that time is an irreversible arrow, a capitalization, a progress. (68–9)

Scientific positivism becomes one of the key discourses of modernity, then, because it asserts its ability to codify the phenomenal world according to rationally legible, often teleologically inclined, schema that break with the supposed irrationality of the past.

This notion of a temporal rupture with the pre-modern also appears clearly within neurology's accounts of itself during this period. For example, for Thomas Trotter in 1807, understanding the nerves required the toppling of anatomical methods for studying the organism:

> Dissections have not forwarded our knowledge of these [nervous] diseases: and indeed when we consider the nature of their symptoms, symptoms flying from one organ to another in an instant, and thought succeeding thought, with the rapidity of lightning, we are the more inclined to think that inspection of dead bodies will not improve our method of cure. (213)

Within its discourses of function, however, neurology also endorsed this notion of historical rupture by re-conceiving of the body as something exposed to a constant influx of new and often increasingly intense stimuli from the external world of technological modernity – an idea that found articulation in broader cultural discourses. In H.H. Munro's short story 'The Peace of Mowsle Barton' (1911), an anonymous traveller comments, on arrival at Paddington Station: "Very bad for our nerves, all this rush and hurry...give me the peace and quiet of the country" (121). As we will see, the neurological body discovered the potential for distress and even trauma in every corner of the modern world, intensifying modernity's understanding of history as founded on conflict, schism and overcoming. Neurology, as this volume will demonstrate, also enhanced the sense of modernity as a product of rupture by creating new ways of understanding the body's internal workings and new experiences for it to process that were represented as having little historical precedent.

So neurology and modernity worked together to create narratives of legibility for previously occluded experiences and structures, registering as 'data' occurrences that had previously been either unnoticed or unavailable. Nevertheless, although neurology assumed a recognizable place within nineteenth-century scientific positivism, forming part of its major narrative of progression, it is vital to recognize that it also created complex versions of subjectivity, and, by implication, relations between individuals and society, which seemed always to exceed these capacities for rational explanation. The drive to press the object world to supply data produced material that exceeded the capacities of the system to interpret it, data that justified the existence of the system by forcing it to extend its inquiries and increase its complexity. Producing and gathering 'data' and presenting that material as its rightful object of study indeed consolidated neurology's discursive power, allowing it to present itself as inhabiting or carving out a new space. But this (mass)-production of subject matter and interpretive discourse also caused the limits of neurology (as with all knowledges) to remain permanently unstable, subject to new paradigms and materials that forced their constant reconfiguration. Neurology was particularly troubled by the tense relationship between its scientific status and the uncontrollable cultural life of its major terms and conceptions, and should be recognized as a hybrid rather than a successfully 'purified' discourse, in Latour's terms. In other words, it was always negotiating between its more centripetal elements and the centrifugal forces with which it remained fascinated: the radical injuries, disorders and diseases found within bodies that led complex cultural lives.

Suggestively like the modern nervous systems it sought to describe, neurology proved itself to be particularly susceptible to absorbing the impressions, sensations and contaminations of the broader cultural discourses in which it was immersed. To read neurology's relationship to modernity and to trace the network of connections by which different cultural and scientific discourses mirrored and motivated one another in their accounts of the human body and mind, is thus to begin to see the complexity of the ways in which particular aspects of the material facts and functions of the phenomenal world become visible at certain historical moments. To read this network of connections, which sometimes runs in unexpected directions, is indeed in one sense to read neurology *against* the putative modernity it seeks to assume for itself. Tracing out the hybrid aspects of neurological discourse rather than its roots in the scientific positivism of modernity, paying attention to the vitality of the network of ideas, fascinations and fantasies of modernity into which it was plugged, demonstrates how both neurology and modernity have never simply been 'modern', in Latour's sense of the world. Such a method also presumes that the natural world of material phenomena is neither sufficiently weak that it is simply there, waiting to be dominated and reformed under humanity's discursive power, nor sufficiently strong to resist completely the cords that bind it to our interests, beliefs, ideas and representations. The idea of tracing a network of connections and opening up the spaces of mediation between what is idea and what is matter, what is preformed or given and what is constructed or built, also resists linear temporality and simple assertions of priority in which one set of circumstances is rendered as obviously producing another. What it allows, instead, is the possibility of reading the complex yet determined relationships between humans, non-humans, things and ideas at particular historical moments. For Latour, the idea of the network, "[m]ore supple than the notion of system, more historical than the notion of structure, more empirical than the notion of complexity", enables the illumination of what he perceives to be a vast excluded middle space of historically determined, knotty and involved relationships (3). And it is the idea that there is a network of mutually interpenetrating connections between neurology and modernity, rather than the notion that either category is simply discursive or, conversely, and equally reductively, a transparent window on to a world of immutable natural functions, that *Neurology and Modernity*, in various ways, explores.

It is perhaps more than coincidence that in 1973, when the anthropologist Clifford Geertz asserted the complex interpenetration between culture and the material world within the social, he turned to neurology for an example of mutual interpenetration. "Our ideas, our values, our acts, even our emotions", he wrote, "are, like our nervous system itself, cultural products – products manufactured, indeed, out of tendencies, capacities, and dispositions with which we were born, but manufactured nonetheless"(51). The chapters in this volume reveal the ways in which neurology was neither fully empiricist – basing its conclusions on tested hypothesis via observation and measurement – nor completely relativist – simply constructing and then expressing the discoveries its practitioners expected to find. As supple and complex as the nervous system it sought to describe, neurology was a network of hybrid discursive practices caught between scientific observation, ideological and cultural predispositions, and sometimes occluded, sometimes explicit, fantasies and desires. It thus saw in its object of concern the multiple modes by which the impressions and sensations of modernity were reflected in somatic and psychological experience; at the same time, however, it received from the network of nervous matter itself new paradigms for understanding human bodies and minds, and their relations to a social world, that were constantly in the process of being reshaped.

In arguing that neurology provided the basis of the self of modernity, and modernity gave rise to the priorities and representational tropes of neurology, this work explores one example of the peculiar dynamic through which ideas achieve stability and visibility – forms of mutual citation by different but interpenetrating discourses. Modernity's most central conviction, for example, the idea of human society as undergoing a process of unceasing change, could insist that neurology had uncovered proof of the immanent and constant nature of mutability in its discovery that the body's 'rhythms' were fundamentally organized via the nerves' internal distribution of their impressions of change. Equally, neurology could invoke modernity's certainty that everyday existence was a matter of bombardment by sensory assaults to underwrite its notion that nerve fibres are subject to fatigue equivalent to that of muscle fibres. Modernity could rely on neurology having proven that every human action was motivated and thus could be explained by judicial and diagnostic practices, whilst neurology could cite as one of its justifications the singularly modern belief that the seemingly chaotic actions of entire populations would reveal deep, merely so far unobserved, patterns if they were scrutinized closely. The notion that rapid business communication was fundamental to making us modern (the impulse behind the invention of the electric telegraph in the 1830s) could consolidate its discursive force through the almost simultaneous discovery, by neurology, that our very viscera work according to the rapid circulation of messages. Neurology in turn justified its perception that the activity of the nerve fibre was a matter of message-transmission rather than fluid-transmission by implicitly citing modernity's inquiries into the speeds of such natural phenomena as light, which was first roughly measured in 1849. This two-way mutual citation permitted two seemingly distinct discourses both to achieve and to maintain the status of knowledge.

Neurology and the self of modernity

Although it is clear that the experience of modernity was mirrored in a common perception of the body's high susceptibility to nervous agitation, the very formation of a number of the constitutive elements of the modern subject can be seen to have strong roots in the assertions and concerns of the new discourse of neurology. Basic ideas of the nerves as unique structures serving as avenues for the will dated back to antiquity. In his collected works, published in 1575, the French 'chyrurgeon' Ambroise Paré (c.1510–90) described the nerves as "the ways and inftruments of the animal fpirit and faculty, as of which thofe fpirits are vehicles" (113). For most historians of neurology, however, it was only with the work of Thomas Willis (1621–75) that the nerves began to be awarded more complex functions, with Willis's *Cerebri Anatome* [*The Anatomy of the Brain*] of 1664 proposing the term 'neurologie' for 'the doctrine of the nerves', as translated into English by Samuel Pordage in 1680 (Feindel 1). Challenging the teachings of William Harvey (1578–1657), who drew on the Galenic doctrine that had been dominant in Europe since the twelfth century, and in which the spirit existed materially in the act of fermenting the blood, Willis proposed a nervous system that "exercised a dominant role in the functioning of the body in general" (Clarke and Jacyna 6). "The Brain is accounted the Chief feat of the Rational Soul," Willis wrote, arguing that it channelled 'animal Spirits' through hollow, porous nerves to commune with the faculties, while the nerves kept the Rational Soul safe from the influence of the lower organs (91, 87–90).[2] Imagining was a wavering of the animal spirits, and remembering was an inflow of spirits from the outer parts of the brain (87–91, 95–6). Willis's idea of a 'nervous juice' distilled from blood in the arteries of the brain both neurologically echoed Harvey's 1628 discovery that (and how) the blood was circulated by the heart, and drew on ancient Greek medical notions of a liquid secreted by the brain. In 1703, John Freind wrote that swollen blood vessels could interrupt "the influx of the Spirits thro' the Nerves" and cause disorders such as indigestion in the organs which those nerves supplied with "liquidum Nervofum" (91, 103). But even these hydraulically conceived nerves were not seen as the body's primary structure. They were seen, for example, as incapable of distributing disorder from one part of the body to another. The brain still reigned over the body in 1733 when George Cheyne explained that "the Intelligent Principle, or Soul, refides fomewhere in the Brain, where all Nerves, or the Inftruments of Senfation terminate, like a Muſician in a finely fram'd and well-tun'd

Organ-Cafe; that thefe Nerves are like Keys, which, being ftruck on or touch'd, convey the Sound and Harmony to this fentient Principle, or Mufician" (4–5). Though there was an ongoing debate about whether the nerves channelled fluid, he continued to assume the existence of "a fubtile, fpiritous, and infinitely elaftick Fluid, which is the Medium of the Intelligent Principle" (88–9). Nervous disorders comprised "almoft one third of the Complaints of the People of Condition in England" (ii) only because, as a result of poor diet, the aristocracy's nerves were too relaxed to pump this fluid efficiently, or because the consistency of the fluid was corrupted, so creating a 'nervous' distemper by locally disrupting movement or sensation (14–5).

In the second half of the eighteenth century, however, neurology began to disrupt this dictatorship of the brain. In 1764, Robert Whyte (*a.k.a.* Whytt, 1714–66) determined in his preface to the second edition of his *Observations on the Nature, Causes and Cure of those Disorders which have been commonly called Nervous, Hypochondriac, or Hysteric* that while a nervous fluid might exist,

> the extreme fmallnefs of the nervous tubes, and the fubtility of that fluid which they contain, make us altogether ignorant of its pecu- liar nature and properties. Nor do we know, certainly, whether this fluid ferves only for the nourishment and fupport of the nerves, or whether it be not the medium by which all their actions are performed (2).

"In reasons on the nature and caufes of nervous disorders", he wrote, "I have endeavoured to avoid uncertain hypothefes; and therefore have had no recourfe to any imaginary flight, repercuffion, difperfion, confufion or jarring conteft of the animal spirits; for whofe exiftence we have only probability" (v). Whyte proposed that, in any case, no fluid carried by so thin a tube as a nerve fibre could be the hydraulic cause of muscle contraction (5–7, 68). It was noted in 1784 that the hypothesis of a 'nervous fluid' was waning (Prochaska 381). In 1798 Everard Home presented the Croonian Lecture to the Royal Society on his experimental and microscopic discovery of proof that nerves were solid (12), in 1830 Charles Bell ridiculed medical doctors who still held to the 'nervous fluid' theory (224), and in 1840 Thomas Laycock attributed to the 1837 volume of Johannes Müller's multi-volume *Elements of Physiology* the establishment of a new orthodoxy that there is no nervous fluid (101).

In 1764, Whyte proposed a new mechanism by which hypotheti- cally solid nerves were seen to operate, a mechanism which functioned according to "a particular and very remarkable ſympathy between ſeveral...organs, by means of which many operations are carried on in a ſound ſtate" (15). Galenic medicine had noticed the property of 'sympathy', seeing vapours ascending from the stomach, for example, as the cause of a headache where no pathology existed in the head. As Whyte detailed, seventeenth-century writers had assumed that sympathy occurred according to their perceived similarities in structure (38). Willis had also explained the place of nerves in disorders of sympathy, through which a sensation in one area of the body might cause a similar sensation in a distant area. For Whyte, however, this property was no longer a mere oddity of nerves which produced pathology. It was a systemic property that underpinned everyday functioning. He pointed out that the nervous system was sophisticated enough to be able to distinguish between different stimuli: tickling the feet causes convulsions, where inflammation or wounding of the feet does not (56). Whyte also proposed that the involuntary motion of a muscle, rather than resulting from any property possessed by the muscle that would work even when severed from a nerve, was produced, as with voluntary action, by the controlling influence of a nerve fibre (5).

This functional re-invention of nerves was impelled further when the mechanistic, anatomical conceptions of the human body that dominated medicine in the late seventeenth and early eighteenth century began to find themselves confronted by new accounts produced by physiology, with its search for functions that might not be apparent from the anatomical arrangement of organs and fibres, towards the end of the eighteenth century. In spite of his 1764 argument that sympathy was fundamental to everyday functioning, Whyte still saw sympathy operating via the brain, as "an immediate conſequence of the disagreeable perception which excites it into action" (74). This tenet was consolidated by the existence of sympathy "between many parts, whoſe nerves have certainly not the ſmallest communication with one another" (42). There was not, for mid-eighteenth-century neurology, "any union or connexion of...nerves" (Whyte 52–3): they were still the subservient tendrils of a "ſentient principle" (Whyte 61) firmly located in the brain – its penny postmen. The last 30 years of the eighteenth century, however, saw this primacy begin to topple. In his *Principles of a Physiology of the Proper Animal Nature of Animal Organisms* (1777), John Augustus Unzer (1727–99) was one of the first to suggest that sympathy did not need the participation of the brain. "The direct nerve-actions of an external impression on the muscles are the same as the direct sentient actions of its external sensation", he wrote, "and can cause the same series of movements which these latter excite by their pleasure and pain", meaning that previous neurologists had easily mistaken the actions of nerves for the actions of sentience. But sentience was not, in fact, needed:

> [E]xternal impressions may...excite a whole chain of apparently volitional acts, without one of them being felt, or any conception whatever excited. Hence an animal may, by external impressions only, perform all the organic and apparently volitional movements necessary to its existence, without having either brain or mind, if its body be so constituted (as is quite possible) that all external impressions on its nerves can produce their direct and indirect nerve-actions, without having to excite material ideas in the brain, or conceptions in the mind, connected therewith. (241)

Consequently, "there are many movements considered to be sentient actions of external sensations only, which are nevertheless, at the same time, direct or indirect nerve-actions of an external impression" (242). Unzer stressed the importance of a distinction between sensibility (where the stimulus is consciously felt) and irritability (where the stimulus has an effect but is not consciously felt), and argued that while the former often accompanied the latter, irritability was the basic property of all nerves (244). The nervous system was thus in charge of movements that could also be volitional, could be just as 'conscious' of the outside world as the volitional mind, but operated much of the time without needing to trouble intention.

For Unzer, "many movements which are or may be sentient actions, result from the direct nerve-actions of external impressions on the muscles; as, when the irritated muscle moves a limb by its contractions, or closes a cavity, or, as in the intestines, causes peristaltic movements and numerous writhings" (240). As these actions required stimuli to occur, they were not the result of automatic nervous control, but the volitional mind was unaware of them nonetheless. In his 1784 *Dissertation on the Functions of the Nervous System*, George Prochaska (1749–1820) asserted that

> nerves possess their own *vis nervosa*, which never had a connection with the brain. The experiments that prove this have long been perfectly well known; namely, that if a nerve be cut or tied, although by these means its connection with the brain be destroyed, it is still able, if irritated, to cause the muscles to contract as if its connection with the brain were entire. (397)

The *vis nervosa* was the 'agency' of the nerves, and Prochaska pointed out that while Albrecht von Haller (1708–77) had proposed the term, Unzer had been the first to show that this agency was not purely volitional (380). Prochaska insisted that "reflexions of sensorial impressions into motor are effected in the sensorium commune itself while the mind is altogether unconscious" (432). He explained that the *vis nervosa* is "ever latent, and exists as a predisposing cause, until another exciting cause, which we term stimulus, is brought to bear". He then divided these stimuli into two types:

> either it is some fluid or solid body applied internally or externally to the nervous system, and termed corporeal, or mechanical stimulus; or else is a mental stimulus present in a portion of the nervous system, and by means of this portion controls the rest of the nervous system, and the rest of the body, as far as it is allowed. Whether this mental stimulus takes placed through a system of occasional causes, or pre-established harmonies... or... by a physical influx, matters little to our object; it is sufficient for us that the soul can excite the nervous system to the performance of certain actions, and this power we call a mental stimulus. (390)

This new autonomy for nerves, in which the commands of the rational soul were now mere 'stimuli' equal to those external and internal mechanical stimuli which caused the *vis nervosa* to activate, was reflected in pathology, with William Cullen explaining in 1783 that "[i]n a certain view, almoſt the whole of the diſeaſes of the human body might be called NERVOUS" (141). Cullen proposed "the title of NEU-ROSES" to denote "all thoſe preternatural affections of ſenſe or motion which are without pyrexia [fever], as a part of the primary diſeaſe; and all thoſe which do not depend upon a topical affection of the organs, but upon a more general affection of the nervous ſyſtem" (142). For him, nerves now operated as a non-conscious brain diffused throughout the body.

Thomas Trotter detailed such diseases in 1807, writing that, for example, men of literary character, from "poring too long over the same subject,... subtract from the body much of that stimulation which is required for many operations in the animal economy", therefore causing the "powers of digestion" to "grow unequal to their office" (39). In turn, the debility and inactivity in the digestive system "re-act on the nervous part of the frame; and the faculties of intellect, as sympathizing in a great degree, with all these highly sensible bowels, are influenced by the general disorder" (39). For Trotter, while the mind might be one origin of disorder in the nerves, nerves could equally cause disorder in the nervous matter of the brain. And while earlier commentators had noted a peculiarly strong sympathetic nervous link between the stomach and the brain, Trotter understood this sympathy not as a single track but as a network of operations that allowed movements in a number of directions, observing that "we find all those viscera, which assist in preparing the chyle, and what is called the assimilation of the food, joined in a circle of nervous communication, of which the stomach is the centre" (223–4). Nerves were now seen as communicating with each other rather than simply running the errands of the brain. Another account of nerves as agents rather than subservient units came from Everard Home, whose turn-of-the-century investigations into the irritability of nerves led him to conclude that "[i]n many diseases, there are symptoms so decidedly confined to the courses of the nervous chords, that an impartial observer would be unable to account for them, in any other way than by supposing them to arise from some action in the nerves themselves" (20). In 1830 Charles Bell wrote of every 'nerve cord' as a discrete organ (20), and this would become a refrain of early nineteenth-century neurology.

The beginning of the nineteenth century also saw the organic functions of the body withdrawn from control by the brain and relocated in the ganglia of the 'vegetative nervous system'. In 1800 Marie François Xavier Bichat (1771–1802) and Samuel Thomas Soemmerring (1755–1830) independently proposed that vegetative nerves were completely autonomous from cerebrospinal nerves: although linked, vegetative nerves (at the time seen as a network of branches from the 'great sympathetic' nerve) had their own independent nervous power source, did not depend for their operation on either the brain or the spinal cord, and had their own ganglia – mistakenly interpreted so far, they noted, as nodes of intersection providing connections between nerve fibres – as their nervous centres.[3] Franz Joseph Gall (1758–1828) and Johann Gaspar Spurzheim's (1776–1832) collaborative work between 1800 and 1815 led the latter to interpret the ganglia as the true origins of all nervous fibres, adding that they "abstract the parts they furnish with nervous energy, from the influence of the will" (22). This was one of the pieces of evidence that moved Spurzheim to interpret "the nervous masses of vegetative life as independent of those of phrenic life [i.e. the brain and spinal cord], in as far as their existence is concerned", in spite of seeing parts of each as devoted to communication with the other (23). By 1830, summing up his papers on nerves to the Royal Society between 1821 and 1829, Charles Bell referred to Bichat's independent vegetative nerves as an entire nervous system, based in the semilunar ganglion and the solar plexus (which he saw as connected to the nerves of the cerebrospinal system only by small twigs linking it to the spinal marrow), and distributed throughout the entire body, "to unite the body into a whole, in the performance of the functions of nutrition, growth, and decay, and whatever is directly necessary to animal existence" (16). The groundwork was being laid for later proposals that the body contained not one nervous system but several.

The idea that nerves possessed not just an agency but a consciousnesses of their own was further endorsed in the first 30 years of the nineteenth century, when the belief in non-cerebral communication between nerves, which Prochaska still saw as a fraught question in 1784, began to be formulated into an idea of a second nervous system. This required a further unseating of the brain, this time from the position of 'origin' of the nerves. In the 1820s, Spurzheim remarked that "[c]omparative anatomy, and acephalic monstrosities among the mammalian and man, furnish incontrovertible proofs of the brain not being the origin of the nervous system at large" (13). Comparative anatomy showed that "very many of the inferior animals have nerves, although they have nothing that may be likened to a brain. Their nervous system, consequently, cannot have had the origin [in the brain] commonly assigned to it by authors" (13). And as 'acephalic monstrosities' proved that parts of the nervous system develop independently in gestation, "[t]he first anatomical principle in regard to the nervous system therefore is, that *it is not an unit, but consists of many essentially different parts, which have their own individual origins, and are mutually in communication*" (14–15, emphasis in original). Spurzheim was also one of the first to assert that even the spinal cord is not an extension of the brain, transmitting messages from above, it is instead a discrete organ formed of nerves that run upwards to the brain from below (31–3). In 1830, Charles Bell argued that in the course of the nervous cord from the muscle of the forearm, "in all this extent, however combined or bound up, it constitutes one organ, and ministers to one function, the direction of the activity of a muscle of the hand or finger" (20). As Edwin Clarke and L.S. Jacyna point out, by the time of the publication of the first volume of Robert Todd and William Bowman's textbook *The Physiological Anatomy and Physiology of Man* in 1845, the Willisian idea that the whole nervous system comprised merely 'feelers', qualitatively different from the brain that sent them out, had been evicted by the common conception, based on comparative anatomical studies, that the brain was merely an upward extension of the spinal cord, a mere "aggregate of gangliform swellings" (246–7; Clarke and Jacyna 31). While for George

Cheyne, in 1733, consciousness, or "the Intelligent Principle" (4), controlled the nervous system, for mid-nineteenth-century neurology the nervous system was a form of consciousness. This notion would soon be further strengthened by the formalization of the theory of evolution, versions of which were already circulating in specialist circles by the 1840s, by Charles Darwin in 1859.

While, in 1837, Benjamin Brodie was still maintaining that sympathy between a damaged part of the body and a distant and undamaged part occurred via "reflection" in the brain (14), from 1832 Marshall Hall had begun to publish discoveries of the nature of the 'reflex arc' which revolutionized understandings of nervous communication. The reflex arc was, he argued, a property of the spinal marrow which turned sensation directly into action without that sensation needing to pass via the brain. Hall had found apparently voluntary actions occurring in a number of decapitated amphibians in response to stimuli (an experiment that he by no means invented), and on this he built a theory of an entire catalogue of reflex arcs, linking sensory and motor nerves distributed to all points in the body, that was built solely into the spinal cord, and which, by sensing both external and internal changes, determined a vast range of bodily processes. These processes included those, such as breathing, which could be taken over by the brain when volition was exercised, but which occurred constantly without requiring the participation of the brain. While the stimuli activating reflex arcs included those also felt by the brain, the brain's awareness of them was not required for reflex action to occur. Indeed, this 'excito-motory' property of the spinal cord, as he proposed it be called in 1837 (*Memoirs* xiv), constantly sensed and acted without the participation of the brain in either. Reflex actions would occur, he pointed out, in a region where the nerves of conscious sensation had been destroyed (*Memoirs* 51).

Hall determined that, across the body, "the reflex function exists as a continuous muscular action, as a power presiding over organs not actually in a state of motion, preserving in some, as the glottis, an open, in others, as the sphincters, a closed from, and in the limbs, a due degree of equilibrium, or balanced muscular action" (*Memoirs* 5–6). More significant than the sympathy observed by Whyte, Unzer and Prochaska, this 'excito-motory' property comprised an entire intermediate nervous system distinct from the volitional nervous system by virtue of its independence from the brain, and distinct from the vegetative nervous system because of this dependence on a constant influx of external and internal stimuli not sensed by the brain. This system, Hall argued, managed internal balance and sphincteric control, in such specific instances as gagging, swallowing and the distending of the cloaca when material is pushed into the intestine, or in the closing of an eyelid when the eye is touched (*Memoirs* 13–14). Its impulses may not have been spontaneous, but neither were its nerves the mere messengers of sensation or voluntary action between nerve endings and brain; instead, nerves themselves undertook a kind of intellection, interpreting stimuli independently of the brain (Figures 1 and 2). As Hall argued, "[t]he cerebral system connects us with the external world in everything that relates to sensation and volition, or mind; the true spinal [i.e. excito-motory] system, in everything that relates to the appropriation of its materials, or their expulsion, – in everything that, in those respects, relates to nutrition and reproduction" (*Memoirs* 71). Hall's "system of excitor nerves, constantly operating in the animal economy, preserving its orifices open, its sphincters closed, and constituting the primum mobile of the important function of respiration" (*Memoirs* 74) described an organism whose proper functioning (including, Hall would later add, circulation and digestion) was ensured by a constant stream of internal and external nervous stimuli (*Synopsis* vii).

Renamed the 'diastaltic nervous system' by Hall in 1850, this new nervous territory was to become the subject of widespread investigation by mid-nineteenth-century neurologists. In 1840 Thomas Laycock stressed that not only did reflex arcs produce secretions and movements in response to exciting causes without producing any conscious sensation, if the brain were to become incidentally conscious of the exciting cause, it would be aware of a much less complex stimulus than that experienced by the excito-motory system (104). Neither were reflexes ever found to be merely automatic muscular contractions. In 1853, Eduard Pflüger showed that if a frog was decapitated and a drop of acid placed on its thigh, it would try to wipe the acid off with the other leg. This was a complex enough reaction in itself, but he also observed that if the same experiment were carried out with the second leg amputated, the frog would bend the first leg to try to remove the acid (Lewes 2:267). The reflex systems of Pflüger's decapitated frogs were adapting to specific circumstances as if they possessed intelligence. This was confirmed by investigations into the speed of reflex impulses which discovered significant delays between the reception of an afferent signal and the activation of the appropriate efferent nerve, so implying a period of spinal cord 'processing'.

For Laycock, reflex investigations also discredited "the common doctrine that the nerves depart from one common centre". Rather, if nerves had anything resembling a centre, a model in which "they

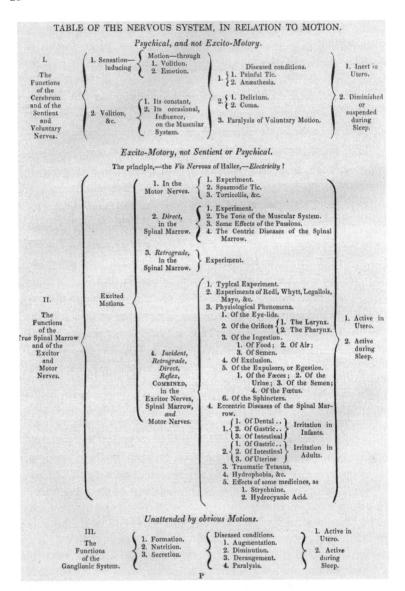

Figure 1 From Marshall Hall, *Memoirs on the Nervous System. Memoir I: On the Reflex Functions of the Medulla Oblongata and Medulla Spinalis*, 105

communicate with several circles, situate upon one common circle [i.e. the brain]", was more appropriate than a hierarchical organization (98). And as Laycock and William Benjamin Carpenter were instrumental in pointing out in the 1840s, because the brain was merely an upward developmental extension of the spinal cord, an extra organ which supplemented the nervous system only in 'higher beings', it too was likely to carry out non-conscious reflex functions (Laycock 105–7; Carpenter 223–4). In the 1850s, Jean Marie Philipeaux and Alfred Vulpian would provide even further evidence of the autonomy of the nerve when they showed that a nerve fibre can live and function if severed from its root in the spinal cord, and in 1861 Augustus Waller announced that he had discovered that this was because the nutritive centre and battery of each nerve is its ganglion (or the spine in the case of nerves with no ganglion) (19–20).

Together these neurological developments produced an account of a body occupied by nerves that (a) were powered independently of the brain, (b) were distributed in several distinct systems in autonomous communication, (c) comprised a sensate and animate mechanism with no dependence on the sensations experienced by the brain and of whose workings the brain was frequently oblivious, (d) required the stimuli of the outside world for the organism to even survive, (e) participated in acts of communication with their most remote fellows and (f) represented, in and despite their automaticity, what might be thought of as the evolutionary origins of the volitional mind. By the time of the publication of George Henry Lewes's *Physiology and Common Life* (1869–70), nerves were perceived as so conscious of the world that it was necessary to distinguish between 'consciousness' and 'consciousness of one's consciousness'. For Lewes, adaptive 'reflexes' showed that it was false to believe "that no sensation can be produced by an impression, unless that impression reach the Brain"; instead, "[t]o have sensations and to be conscious of sensations is one and the same thing. To *have* a sensation and to *know* that we have it, are two things". In the spinal cord "consciousness may, and often does, exist without knowledge" of this consciousness in the brain (2:48, emphasis in original). The sophistication of these 'reflexes' showed that "we may properly say that there can be unconscious thinking, and unconscious sensation" (2:194).

Paradoxically, however, although the period of modernity saw an unseating of the brain as an organ in control of the body, the notion of a nervous network assembled from various reasonably autonomous units that were integrated into a functional whole also motivated a new attention to the specific functions of the brain (see Clarke and

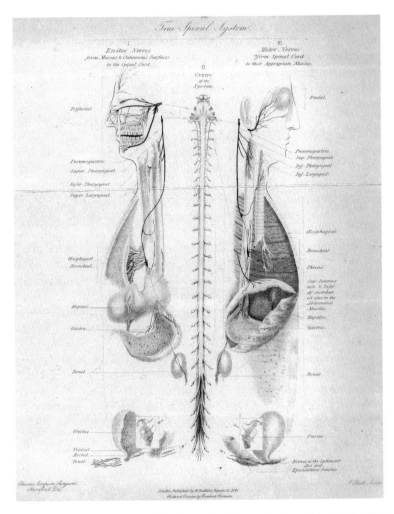

Figure 2   Marshall Hall, *New Memoir on the Nervous System*, Plate II: The True Spinal System

Jacyna 212–20). Although early neurological claims that the brain was the sole origin and centre of the nervous system had been jettisoned by the middle of the nineteenth century, the brain remained central to accounts of such 'higher' functions as thought and speech, and also maintained a significant position within the functioning of even basic motor acts. Neurologists consistently failed to establish any anatomical distinctiveness of the brain; nevertheless, there remained a persistent interest in determining its importance and singularity, even as the brain found itself increasingly impelled and often alarmingly harassed by a nervous system that paid little heed to an alleged separation from the *cogito*. In the eighteenth century, Albrecht von Haller had proposed a unitary theory of brain, in which all of its areas were integrated in their function and contributed to its common activity, but this view was challenged decisively by Franz Josef Gall. In 1791 Gall published an account of 'organology', determining that the mental powers of the subject were constructed by separate psychical and moral faculties, each possessed of its own 'organ' (Spurzheim called the brain "a collection of many peculiar instruments" [106]), specifically located on the surface of the brain. Spurzheim's better-known term 'phrenology' built upon Gall's theorization that the development, or not, of these 'organs' was linked to faculties and could be analyzed by examining the external contours of the skull. But where Gall remained sceptical as to whether any but a handful of subjects possessed skulls that gave a sense of the 'organs' underneath, Spurzheim famously believed that the method could be applied universally, thus decisively inserting a version of localization theories of brain function into popular cultural discourses of the nineteenth century. Robert Verity's 1836 account of nervous system in modern civilization included the assertion that phrenological analysis of portraits of historic leaders could show the mental deficiencies that had caused their failings (130).

Neurology therefore provided several major components of modernity's particular conception of somatic and psychological experience. First of all, it dispersed the origins of the body's functions away from the brain, proposing, instead of a single dictating and presiding origin, a republic of component functional units. These units even included controlling centres wholly distinct from the brain: second and third brains. This established the principle that instead of the nerves and the body being antagonists, the body was nervous in its entirety. Nerves no longer insulated the mind from the body; instead, nerves now comprised governing systems in need, if anything, of insulation from the mind. Neurology indeed distributed consciousness throughout the nervous system, suggesting, as in the case of Lewes's distinction between 'consciousness' and 'consciousness of consciousness', the 'invention' of an unconscious mind long before the appearance of psychoanalytic accounts describing a functional entity called 'the unconscious'. Even the most seemingly unmotivated of actions, this doctrine held, could now be explained as fully determined, even if those determinations were difficult to decode. Neurology also joined the body's processes to external and internal stimuli, by opening up, between volitional control and automatic function (both independent of stimuli), a widening realm of neurologically governed internal activities which depended on stimuli to occur. The self could not be seen as capable of detachment from the world if the entire nervous system, including the brain, required the 'interference' of immediate surroundings to function. If the body was, by its nature, functionally connected to its environment, the modern neurological subject needed to be regarded as fundamentally attached to a world by which it could be both nourished and harried.

Advances in the understanding of neurological functioning also contributed to the idea that nervous disorder was common, distributed throughout the body and the populace as a whole. When George Cheyne listed his eleven "Signs and Symptoms of a too relaxed, loofe and tender State of Nerves" (99) in 1733, nervousness seemed endemic to the body:

> Those who Stutter, Stammer, have a great Difficulty of Utterance, fpeak very Low, lofe their Voice without catching Cold, grow Dumb, Deaf, or Blind, without an Accident or an Acute Diftemper; are quick, prompt, and paffionate; are all of weak Nerves; have a great Degree of Sensibility; are quick Thinkers, feel Pleafure or Pain the moft readily, and are of moft lively Imagination. (104–5)

The rest of this list included such physical disorders as a weak pulse, loose and yielding flesh, excessive sweating, salivation and discharges from the eyes, all of which were caused by "weak, loofe, and feeble or relax'd Nerves", in Cheyne's terms (99). By contrast, by 1909, Annie Payson Call, in her popular advice manual *Nerves and Common Sense*, was asking women "what could keep one in nervous illness more entirely than... [the] deep interior strain which is necessary to... [the] external appearance of placidity"? (223) This would seem to suggest that the early eighteenth century was *more* nervous than the beginning of the twentieth century, when a nervous disorder became a disorder solely of the nerves. However, by locating consciousness in every nerve cell, nineteenth-century neurology had remodelled the nerves as reflections of mental states. As such, nervous disorder could now occur as a consequence of even the smallest psychological changes: it no longer required anatomical pathology to occur. Accordingly, nervous disintegration lurked as an almost constant threat within both modern mind and body.

Neurology also continued to provide the period of historical modernity with new conceptions of how the self might be structured and organized. Major instances of this were the neuron doctrine and the account of neural plasticity. Following the publication of Theodor Schwann's *Microscopical Researches into the Accordance in the Structure and Growth of Animals and Plants* in 1839, the discipline of histology located the basis of the body's tissue types within their cellular structure. However, because "it could not be seen whether the long and thin branches arising from nerve cells and nerve fibers have definite terminations, or whether they run together with neighbouring thin branches to form a continuous network" (Shepherd 4), there was significant doubt over whether nerves were cells. While some used the term over the ensuing half-century (see Waller 9), conclusive proof of the existence of distinct nerve cells was not discovered until the end of the 1880s (see Shepherd 3–4). The *fin-de-siècle* consequently saw a number of investigations into how each of these now undeniably discrete units could receive and pass on nervous impulses, and contemporary studies by Charles Sherrington (1857–1952) and Santiago Ramón y Cajal (1853–1934) established some of the earliest accounts of 'synaptic firing', accounts that confirmed and popularized the image of the body as suffused with an intricate one-directional electrical circuit, and generated, in their description of interneuronal communication, nervous functions that were particularly susceptible to misfiring under the influence of fatigue (Rapport 104).

Nineteenth-century neurology's accounts of the plasticity of body's governing system also contributed to a broader restaging of the ancient debate over whether organisms developed according to their preformed natures or whether they were affected by epigenetic factors such as experience. For Trotter in 1807, for example, the hardiness of the working classes derived from the effects of hard work on their nerves. "How soon", he asked, "would the morning dram of a Billingsgate fish-wife, destroy one of our high bred women of fashion?" (18) In 1837 Benjamin Brodie wrote that "the construction of the nervous system, at the period when growth is concluded, may not be the same in all individuals, and that an imperfect development of it may lay the foundation of all the aggravated hysterical affections" (70). For him poor development of the nervous system led to "an insufficient generation of nervous energy" (72). In 1839 Robert Verity argued that "individuals, in fact, undergo modification in the constituent tissues of their bodies, according to the particular kind of activity which governs and habitually prevails" (2). Verity even believed that life expectancy was greatest in towns because the city was "where the greatest amount of nervous energy is in action; where the individual is habitually immersed in the higher kinds of vital excitement consequent upon intellectual and political pursuits" (113). But as biology began to propose the components of the theory of evolution in the 1840s, accounts of plasticity moved from discovering neurological changes that seemed to echo the division between the 'races', to finding explanations of personality. By the end of the nineteenth century, this account was sufficiently commonplace to appear in popular fiction. In 1895, in Wells's 'The Sad Story of a Dramatic Critic', Egbert Craddock Cummins tells how he began to imitate the gestural stage acting he has been ridiculing in his role as a critic:

> [n]ight after night of concentrated attention to the conventional attitudes and intonation of the English stage was gradually affecting my speech and carriage. I was giving way to the infection of sympathetic imitation. Night after night my plastic nervous system took the print of some new amazing gesture, some new emotional exaggeration – and retained it. A kind of theatrical veneer threatened to plate over and obliterate my private individuality altogether. (246)

By the end of the nineteenth century, then, the nervous self was emerging as both part of one's very nature and a system that appeared to be both creatively and threateningly malleable.

Modernity and the neurological self

As will have become clear, no analysis of neurology will find that it emerged decisively within an easily defined period. Stanley Finger traces the history of neurology back to Galen, whilst Edwin Clarke and L. S. Jacyna authoritatively claim that "by 1850 the foundations of modern neuroscience had been laid" (1).[4] Nevertheless, the period of modernity is significant for the exponential growth and consolidation of neurological discourses, discourses which tended to figure themselves as breaking new ground, as offering radically new conceptions of the human. Modernity was thus neurology's moment of condensation; it is the moment when nervousness assumed a dominant cultural life of its own. This moment, when neurological notions of the self began to migrate across many different discourses, can be traced through the obsessive cultural attention afforded to the adjective 'nervous', which underwent a significant shift in meaning during this period. As a noun, 'nerve' derived from the Greek for 'sinew' or 'string', and it had long possessed an adjectival use connoting 'strength', with 'nervous' meaning 'sinewy' or 'well strung'. In the 1691 translation of Ambroise Paré's 1575 collection of medical works, the word 'nervous' was used to describe the substance of an organ which dilates and contracts (83), and in the early modern period, partly because of this perceived contractility, 'nervous' continued to imply strength. Edward Philips's 1658 dictionary of difficult words defined 'Nerve' as "a ſinew, alſo by metaphor, force or ſtrength of body" and noted that 'Nervosity' was "metaphorically taken for ſtrength or vigour". In 1735 'nervous' was still defined as "ſinewy, ſtrong, luſty; alſo in an Argument ſolid and weighty" (Defoe).

The first register of a different meaning for the adjective 'nervous', as W.F. Bynum notes, was Samuel Johnson's 1755 *Dictionary*, in which Johnson recorded, alongside the older meanings – "well ſtrung; ſtrong; vigorous" and "[r]elating to the nerves" – a new meaning "[i]n medical cant": "Having weak or diſeaſed nerves", citing its use by 'Cheney' (presumably George Cheyne's treatise on nervous disorders, *The English Malady* [1733]) (Bynum 91). However, the fact that those successors who repeated Johnson's definitions did not repeat his description of the last meaning as mere medical cant does not mean that it was now in common usage (e.g. Sheridan, Walker 1791). The new meaning was not used significantly enough to be recorded at all in John Walker's 1775 dictionary. Even in 1819, although one dictionary defined 'nervous' almost verbatim from Johnson as "Well strung, strong, vigorous; relating to the nerves; having weak or diseased nerves", 'nervy' was still defined, as it had been for Johnson, as meaning "Strong, vigorous" (Walker 1819, 351). But in one of the first departures from Johnson, Charles Richardson's 1839 dictionary defined 'nervy' as synonymous with 'nervous', meaning both "well strung, strong, vigorous, powerful" but "also sensitive in the nerves, sc. to excess, and cons. weak, debilitated, diseased in them" (537). Noah Webster's first dictionary of American Standard English in 1828 also noted that in addition to meaning "Strong; vigorous", "Pertaining to the nerves", and "Possessing or manifesting vigor of mind", 'nervous' had a fourth, at the time colloquial, meaning: "Having the nerves affected". The specialist medical 'cant', with its connotations of constitutional weakness, had begun to expand into a more common connotation of temporary agitation. In 1849 'nervous' was defined as "[r]elating to the nerves; full of nerves, well strung; strong, vigorous; in a common colloquial sense, weak in the nerves, and hence, apprehensive, agitated by trifles" (Smart 407). Joseph Worcester's American dictionary of 1859 gave a detailed description of this new meaning: "Having weak or diseases nerves; easily agitated or excited; irritable; timid; fearful". "Well strung; sinewy; strong; vigorous" was just a separate meaning amongst five (also including 'full of nerves', 'of or pertaining to the nerves' and 'forcible of style') (959).

The first half of the nineteenth century saw a consolidation of popular notions of nervous agitation sufficient to introduce into common usage a meaning that was precisely the opposite of that which had been in operation before the 1840s. Even though the old meanings still obtained, the new meaning widened from the temporary state of having one's nerves agitated, to a permanent state of liability to such temporary agitation. In 1893 'nervous' was defined as "having the nerves easily excited or weak" alongside the earlier meaning: "having nerve: sinewy: strong: vigorous" (Findlater 336). By the time the L–N volume of the first *New English Dictionary on Historical Principles* (later the *Oxford English Dictionary*) was being compiled in earnest between 1879 and 1907, the older meanings were considered either obsolete or rare, leaving just those in which 'nervous' meant 'with much energy' (particularly in reference to written arguments) and the contemporary meanings: "suffering from disorder of the nerves; also, excitable, easily agitated, timid" (earliest citations in 1763 and 1812); "characterized or accompanied by agitation of the nerves" (earliest citation in 1797); and "agitating to the nerves" (earliest citations in 1775 and 1834). It charted 'nervously' as connoting strength from 1641 and implying agitation only from 1838, and it charted 'nervousness' as connoting strength from 1727 and signifying agitation only from 1798 (vol. 6 [1908], 96–7). Modernity was sufficiently engaged with the idea that the body was permeated by nerves whose variability determined human existence and experience that the second half of the nineteenth century saw the general demise of earlier meanings that implied strength and robustness.

If neurology seemed to affect dominant conceptions of what it meant to be a modern subject as ideas about the functioning of the mind and body seeped from science into broader cultural discourses, it is equally true that major aspects of the neurological self can also be seen to follow from the conceptions and characteristics of modernity. The transition from hollow to solid nerves in neurology, for example, might be read as a corollary of a partial transition, in ideas about the body's transmitted substance, from energy to data. Nerves were still seen to use energy: Benjamin Brodie, in his *Lectures Illustrative of Certain Local Nervous Affections* (1837), was one of the first to use the term "nervous energy" as a replacement for jettisoned ideas of a nervous fluid (72). And their ability to work was also deemed to be determined by variations in this supply, but they were no longer seen to circulate energy to other organs. In 1885, Michael Foster defined nerves as "strands of irritable protoplasm whose

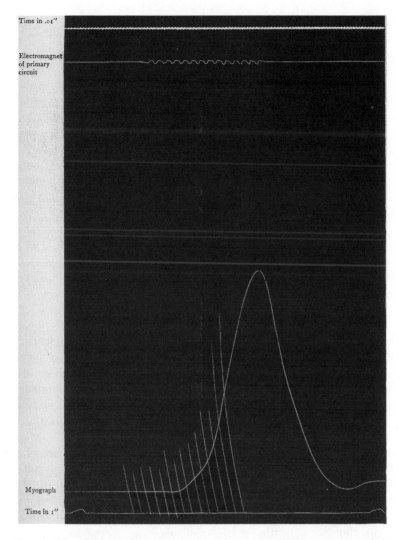

Figure 3 The 'flexion-reflex' observed as reflex contraction (excitation) of the flexor muscle of the knee of a dog. From Charles Sherrington, *The Integrative Action of the Nervous System*, 94

energy is not spent in movement, but wholly given up to the rapid and easy transmission of molecular vibrations" (15). This meant that nerves had to be recognized as both belonging to the same plane as the rest of the body (they were not the channels of animal spirits) *and* of a distinct order (they were not a mere sibling to the blood vessel, carrying instead a '*vis nervosa*' that could move much faster than any fluid). Reconceived as an impulse towards the middle of the nineteenth century, the nerve 'signal' could be, and was, timed (see Figure 3). The German physicist and physiologist Hermann von Helmholtz (1821–1894) 'clocked' human nervous impulses at between 35 and 45 metres per second in 1867 (McKendrick 7; Debru 471). Nerves were measured on, and understood in terms of, an axis – time – the representability of which had been established by the new speeds and frequencies achieved by modernity's pumps, springs, joists, derricks, engines, factories, work crews, transportation networks and communications links (see Doane 4). And as the discourses of a technological modernity achieved dominance, so a model of the body appropriate to its priorities came into being. The body was found to be regulated by limited energies rather than, as in the humoural model, a balance of theoretically unlimited forces.

Data-transmitting nerves also invoked a conception of the human body as comprising relationships between even the lowliest of its component elements. In his *The Integrative Action of the Nervous System* (1906), Charles Sherrington wrote that "[i]n the multicellular animal, especially for those higher reactions which constitute its behaviour as a social unit in the natural economy, it is nervous reaction which *par excellence* integrates it, welds it together from its components, and constitutes it from a mere collection of organs an animal individual" (2). Much changed from the messengers of a dictatorship of the brain described by Willis in 1664; by 1906 the nervous system appeared as the public sphere of a body made up of many semi-autonomous private units. This followed from the modern business principle that the state functioned as a non-centralized network, an economy, for which government by secular bureaucracy rather than land-owners was appropriate.

For Jonathan Crary, who notes neurology's role in generating the discourses of modernity, "a vast range of modernizing projects in the West well into the twentieth century, and more specifically a new inventory of the capacities and functions of an attentive observer" would not have occurred were it not for the neurological proposal, widely familiar by the 1880s, that the whole of the human nervous system possessed, and was constructed according to, reflex functions (164). As stimuli that could not even be felt were seen to be effecting reactions throughout the body, and as these reflexes occurred not just in the spinal marrow but also in the brain, thinking came to be modelled as dependent, like the body's functions, on constant stimuli (see, for example, Lewes, vol. 2, 192), and stimuli in turn produced not just mental action but a movement of nervous energy across the entire nervous system (Crary 167–9). Significantly, the major neurological principle of total-body 'thrill', in strict contrast to the affective sympathy central to the eighteenth-century notion of aesthetic sensibility, found a parallel, perhaps even its foundation, in the unprecedented growth of cities in the nineteenth century. When the number of western citizens living in urban areas exceeded those living elsewhere, the continuous and unrelenting sensory input experienced in modern city centres was reinvented as a norm.[5]

Alongside the putative demands of a technologized, urban modernity upon a reconfigured nervous system, a major discursive shift inaugurated by the institutionalization of Darwinian theories of evolution and natural selection produced a new emphasis within neurology on determining and categorizing the different stages of nervous development seemingly to be found between humans and animals, within human groups, and even within the nervous system itself. In seeking physiological explanations of psychological phenomena and traits, Herbert Spencer (1820–1903), for example, adapted Darwinian ideas of natural selection to Lamarckian theories of use-inheritance (the idea that organs are enhanced or diminished by use or disuse with the resultant changes transmitted to subsequent generations), to establish the principle that the emergence of volitional rather than automatic action was the *telos* of the evolutionary process in humans. In the 1870 edition of Spencer's *Principles of Psychology*, it was indeed the problematic notion of the 'development' of the nervous system that provided a means for distinguishing between "different grades of men":

> those having well-developed nervous systems will display a relatively-marked premeditation...a greater tendency to suspense of judgements and an easier modification of judgements that have been formed. Those having nervous systems less developed, with fewer and simpler sets of connexions among their plexuses, will show less of hesitation – will be prone to premature conclusions that are difficult to change. (1:581)

The evolutionary development of the nervous system became, for Spencer, the keenest marker of the civilization of humanity itself.

When writing on aphasic symptoms of brain injury in 1879, John Hughlings Jackson (1835–1911) also explicitly aligned neurological function with an evolutionary schema drawn from Spencer. Jackson conceived of three levels in the evolution of the central nervous system: the lowest comprised the spinal cord, the medulla oblongata and pons Varolli; the middle included the basal ganglia and the cerebral cortex before and behind the central sulcus; the highest consisted of the prefrontal and occipital lobes of the brain. Even in his account of the operation of higher functions, such as language production, Jackson forged hierarchical distinctions drawn from evolutionary models. Jackson observed that the aphasic symptom, so often the result of brain damage, was very rarely a complete wordlessness; rather, it was a form of propositional speechlessness, with the emotional (the more automatic) language being preserved where there was a loss of intellectual (the more voluntary) qualities of speech (152). For Jackson, brain damage thus explicitly represented a process of dissolution, somewhat continuous with a threat of degeneration: "there are degrees of loss of the latest acquirements with conservation of the earlier, especially of the inherited, acquirements: in each there is Dissolution, using the term as Spencer does, as the opposite of Evolution" (149). As many other neurologists of the period affirmed, the nervous system and brain function became sites where the markers of evolutionary development could be observed and their implications for the taxonomies of modernity anxiously staged.

Neurology's systems also shared modernity's particular interest in productivism, or the ways in which the human body, society and nature were linked by what Anson Rabinbach has called the "primacy and identity of all productive activity, whether of laborers, of machines, or of natural forces" (3). In 1847, Helmholtz established the universal law of the conservation of energy. A pioneer of thermodynamics, he asserted that there was a single, universal, infinitely transferable energy basic to all nature that could neither be added to nor destroyed. At almost the same moment, however, Rudolf Clausius (1822–88) identified the second law of thermodynamics: the irreversible decline of energy according to the action of entropy. As Rabinbach explains, this law asserted that "in any isolated system the transfer of energy from a warmer to a colder body is accompanied by the decrease in total available energy", implying that all systems tended naturally to move towards a disordered state and that force inevitably dissipates (3). In one sense, the seeming exponential growth in the increasingly disordered nervous systems of the modern subject both expressed and gave an evidential force to the new idea of the inescapability of dissolution and exhaustion found in physics. Neurology had begun to explain nerves as consumers of power when it started using the term 'nervous energy' in the late 1830s (e.g. Brodie 72); consequently, nervous degeneration seemed a scientific inevitability. But neurology did more than simply reflect the dissipation of forces that modernity came to recognize as inescapable. Modernity's desire to regulate and rationalize in the face of what it perceived to be the inevitable disintegration of systems of order, can also be found in neurology's general positivism, in its desire to accord the nervous body rationally legible functional systems. As a discipline that simultaneously discovered chaos (or at least unreadable complexity) in the natural world *and* invented innovative methods of treating this chaos as 'phenomena' – as observable parts of a hypothetically larger unknown order – neurology became modernity's representative science of the body.

Nervous states

The establishment of psychology as a distinct discipline at the end of the nineteenth century, in particular William James's description of consciousness as a river or stream that first appeared in an 1884 essay (2), has been seen as one of the major impulses behind the emergence of (at least) the literary mode we now call modernism in the 30 years after 1890. But it can be argued that such notions were not the exclusive province of psychology, for neurophysiology conceived of consciousness in a similar way. In work on the functions of the brain added to the fifth edition of Michael Foster's *Text Book of Physiology*, published in 1890, Foster explained that

> All life long the never ceasing changes of the external world continually break as waves on the peripheral endings of the afferent nerves, all life-long nervous impulses, now more now fewer, are continually sweeping inwards towards the centre; and the nervous metabolism, which is the basis of nervous action, must be at least as largely dependent on these influences from without, as on the mere chemical supply furnished by the blood. (1116)

Consequently, Foster explained,

> we must regard the supereminent activity of the cortex and the characters of the processes taking place in it as due not so much to the

intrinsic chemical nature of the nervous substance which is built up into the cortical grey matter as to the fact that impulses are continually *streaming* into it from all parts of the body, that almost all influences brought to bear on the body make themselves felt by it. (1117, emphasis added)

Foster even went so far as to claim that if the cortex, "its ordinary nutritive supply remaining as before, . . . were cut adrift from afferent impulses of all kinds" – i.e. if the stream of sensory input was stemmed – "volitional and other psychical processes would soon come to a standstill and consciousness vanish" (1117). The famous stream of consciousness of the 'I' found in the new modernist novel, and first identified by May Sinclair in 1918 (444), can thus be viewed, in one sense, as part of the nervous system's 'natural' response to external stimuli.

Although such material can be used to widen our understanding of the discourses which contributed to representations and understandings of aesthetic technique, *Neurology and Modernity* approaches neurology as more than mere 'context' behind the supposedly more important business of aesthetic production. These chapters do offer up new readings of particular works of art, but, as a broader exercise in cultural history, they survey a wide range of discursive products – a spectrum running from such ephemeral creations as gossip, through the seemingly authoritative solidity of scientific treatises, to such resolutely material objects as train gauges. This work tackles the historicity of government advice manuals, receipt books, mutual slanders, self-help works, science fiction, literary fiction, public health warnings, philosophy, spiritualist treatises, government reports and military tribunals. It also analyzes such broad human products and practices as understandings of the ways in which the body is (supposedly) sexed, the nature of the training of children that ought to go on in the home, and the body's methods of habituation and adaptation. This work is finally as labile as the discipline it treats. For neurology was never a grand project or a discipline into which the raw human subject was plugged; instead, it constituted a nexus in which new and disparate ideas came together to produce knowledge. Neurology indeed always assumed a role in broader cultural concerns, ideals and antipathies, taking its part in diatribes against quackery, injunctions to adopt certain regimes of self-care, and assertions about the existence and classifications of race. So, like much contemporary medical history, the studies in this volume illuminate not simply a history of medical discoveries and methods. They study instead the complex interactions between academic discipline, clinical practice and broader cultural concerns.

The following chapters show, through a range of studies arranged in broadly chronological order, the diverse processes by which neurology and modernity constructed and reflected one another. For Michael K. House, early nineteenth-century prototypes for the unconscious are revealed to be a consequence of the simultaneous urges to combat the reduction of the mind to the brain and to discover the material basis of the irrational. Andrew Shail shows how neurology introduced an unprecedented degree of purposive function into models of the sexed body. Looking at changing understandings of the cause of menstruation, he highlights the ways that the neurological body undermined the pathologization of femaleness. For Hisao Ishizuka, the notion of 'nervous exhaustion' in the middle of the nineteenth century was a product of the bourgeois colonization of the concept of honourable work. Ishizuka explains how the nineteenth-century version of the 'English malady' – nervous dyspepsia or biliousness – was seen particularly to afflict middle class 'brain workers'. As a disease that resulted from hard but non-manual labour, it was a form of nervousness that nevertheless signified industriousness, and was therefore imbricated with a bourgeois protestant work ethic.

Jane F. Thrailkill explores the relationship between neurology and modernity by describing how the railway accident, that epitome of the assaults of the modern world, became an exemplary trauma in its shock to the integrity of the masculine body. She argues that it was the terror-stricken man, rather than the hysterical woman, who thus became the foundational figure for modern trauma theory. For Aura Satz, the neurological preoccupation with the reality of the invisible and immaterial followed the broader cultural preoccupations of modernity. Concentrating on mid- to late-nineteenth-century articulations of phantom limb phenomena, she discovers suggestive resonances between new neurological ideas of proprioception and the body-schema, and the rhetoric and tropes of spiritualism. By using the discursive force of spiritualism, neurology was able to imagine the phantom limb into existence in a way that matched the strange phenomena emerging from patients' traumatic accounts of their own experience. George Rousseau's work also shows how the new definition of paranoia introduced in modernity was based on the model of a body that inscribed experience through sensation onto the nerves. It thus internalized the shocks and exposures precipitated by modernity within the very core of subjectivity itself.

Shelley Trower reads the relationship between neurology and modernity by suggesting that nervous energy was modelled after a parade of vibrating forces being discovered outside of bodies, even within the ether, at the time. The vibrations of the world of modernity became both an assault upon the body's proper functioning and a force that could be harnessed therapeutically to retrain and reform the excesses of vibrating corporeality. For Michael Angelo Tata, the nosology of neurological and psychological disorders within modernity created the conditions for the clinical subject to own, to stand in for, specific pathologies. Using the descriptions of the neurological paranoia of Judge Schreber, Tata reveals how the case study, tied relentlessly to the subject within whom each pathology appeared, produced the conditions for a culture fascinated with traumatized, scandalous and recognizable personalities who emerged and were exchanged as commodities within modernity. For Vike Martina Plock, accounts of the modern self in the work of Edith Wharton are revealed as staging a tense yet typically modern encounter between newly materialized, neurophysiological models of brain and nerve functioning, and an awareness of the potential conceptual and analytical limits of modern science for exploring subjective interiority.

Laura Salisbury uses material that emerges from the treatment of aphasic soldiers following the First World War to illuminate how language came to be regarded as dependent on an array of complex involvements between the human subject and the material world. As these new conceptions of language reconfigured meaning as emerging from processes of interpretation rather than apprehension, the neurological subject of modernity thus appeared as constantly struggling to become legible to itself. For Jessica Meyer, the relationship between neurology and modernity can be seen in the symptom of shell shock, which, like the self-inflicted wound, occupied an ambivalent position between medical disability and military offence that demanded both treatment and punishment. As such, these 'wounds' articulated a tension between what was deemed to be psychological and what was somatic, what lay within conscious control and what remained unconscious, that appeared both within the functioning of the subject of modernity and within the body of the military as a state apparatus.

Jean Walton proposes that the idea of the independent thinking of the enteric nervous system or 'second brain' was vital in generating our modern notion of habit, and was structurally implicated within modernity's conception of time as a surplus that could be tied to capitalist theories of production. Finally, for Melissa M. Littlefield, multiple endeavours to find the consciousness-specific substance of thought, a 'mind particle', drew authority from modernity's obsession with the production of material explanations of mental functioning. As such, the investigation into the materiality of thought constituted a major disciplinary alternative to the discourse of psychoanalysis, long regarded as modernism's scientific counterpart. As Littlefield and the other authors in this volume show, alongside the new disciplines of experimental psychology and psychoanalysis, neurology emerges to become both modernity's reflection and one of its explicating, authorizing discourses.

Notes

1. For an account of this see Bynum (96).
2. Robert Martensen makes clear the influence of the anxieties of the Restoration establishment of 1660–80 on this new doctrine (22–4).
3. See Clarke and Jacyna (30) and Spurzheim (11).
4. George Rousseau's work has also offered a path-breaking account of the mutually constitutive relations between neurology and culture in the eighteenth century.
5. As a percentage of the total population, the urban population of England and Wales was first noticed to exceed 50 per cent in the 1851 census (which defined a town as inhabited by over 2,500 people) (Law 125).

Works cited

Allen, Grant. 'An African Millionaire.' *Strand* 11 (June 1896): 659–67.
Baudelaire, Charles. 'The Painter of Modern Life.' *Selected Writings on Art and Artists*. Trans. P. E. Chavert. Cambridge: Cambridge UP, 1981. 390–435.
Bell, Charles. *A Series of Engravings, Explaining the Course of the Nerves*. London: Longman & Rees, 1803.
——. *The Nervous System of the Human Body*. London: Longman, Rees, Orme, Brown & Green, 1830.
Bennett, Arnold. *The Old Wives' Tale*. London: Penguin, 1954.
Beresford, J.D. *The Hampdenshire Wonder*. London: Penguin, 1937.
Bitro-Phosphate. 'Nerve Drugging Must Stop.' Advertisement. *Daily Mail* 6356 (16 August 1916): 7.
Bouillard, Jean Baptiste. 'Recherches cliniques propres a demontrer que le perte de la parole correspond a la lesion des lobules anterieurs du cerveau, et a confimer l'opinion de M. GALL, sur le siege ge l'organe du langue articule.' *Archives Generales de Medicine* 8 (1825): 25–45.
Brodie, Benjamin C. *Lectures Illustrative of Certain Local Nervous Affections*. London: Longman, Rees, Orme, Brown, Green & Longman, 1837.
Buchan, John. *Prester John*. London: Penguin, 1956.
Bynum, W. F. 'The nervous patient in eighteenth and nineteenth-century Britain: the psychiatric origins of British neurology.' *The Anatomy of Madness*. Ed. Roy Porter, W.F. Bynum and M. Shepherd. London: Tavistock, 1985. 89–102.

Carpenter, William Benjamin. *Principles of Human Physiology.* 2nd ed. London: John Churchill, 1844.

Cheyne, George. *The English Malady, or, A Treatise of Nervous Diseases of all Kinds.* London, 1733.

Clarke, Edwin, and L.S. Jacyna. *Nineteenth-Century Origins of Neuroscientific Concepts.* London: California UP, 1987.

Crary, Jonathan. *Suspensions of Perception: Attention, Spectacle and Modern Culture.* London: MIT Press, 1999.

Cullen, William. *First Lines of the Practice of Physic.* Vol. 3. 1783. 4 Vols. Edinburgh, 1796.

Debru, Claude. 'Helmholtz and the Psychophysiology of Time.' *Science in Context* 14.3 (2001): 471–92.

Defoe, B.N. *A Compleat Engliſh Dictionary.* Westminster, 1735.

Doane, Mary Ann. *The Emergence of Cinematic Time.* London: Harvard UP, 2002.

Eastwick, Egerton. 'One Season.' *Strand* 11 (March 1896): 281–91.

Feindel, William. 'The beginnings of neurology: Thomas Willis and his circle of friends.' *A Short History of Neurology: The British Contribution 1660–1910.* Ed. F. Clifford Rose. Oxford: Butterworth-Heinemann, 1999. 1–18.

Findlater, Andrew, ed. *Chambers's Etymological Dictionary of the English Language.* London: Chambers, 1893.

Finger, Stanley. *Origins of Neuroscience: A History of Explorations into Brain Functions.* Oxford: Oxford UP, 1994.

Foster, Michael. *A Textbook of Physiology.* 5th ed. 2 Vols. London: Macmillan, 1888–90. Book 3. 1890.

——. 'Physiology.' *Encyclopaedia Britannica.* 9th ed. 25 Vols. Edinburgh: Adam & Charles Black, 1875–1889. Vol. 19 (1885): 8–23.

Freind, John. *Emmenologia.* 1703. Trans. Thomas Dale. 2nd ed. London, 1752.

Galsworthy, John. *The Man of Property.* London: Penguin, 1954.

Geertz, Cifford. *The Interpretations of Cultures.* New York: Basic, 1973.

Hall, Marshall. *Memoirs on the Nervous System.* London: Sherwood, Gilbert & Piper, 1837.

——. *New Memoir on the Nervous System.* London: Hippolyte Baillière, 1843.

——. *Synopsis of the Diastaltic Nervous System: or the System of the Spinal Marrow, and its Reflex Arcs; as the Nervous Agent in All the Functions of Ingestion and of Egestion in the Animal Economy.* London: Joseph Mallett, 1850.

Harvey, Robert. *De Motu Cordis.* 1628. *Anatomical Exercises Concerning the Motion of the Heart and Blood in Living Creatures.* London: Richard Lowndes, 1653.

Home, Everard. *Experiments and Observations upon the Structure of Nerves.* Croonian Lecture to the Royal Society. 8 November 1798.

Hughlings Jackson, John. 'On Affections of Speech from Disease of the Brain.' 1879 *Reader in the History of Aphasia: From Franz Gall to Norman Geschwind.* Ed. Paul Eling. Amsterdam: John Benjamins, 1994. 145–67.

James, William. 'On Some Omissions of Introspective Psychology.' *Mind* 9.33 (January 1884): 1–26.

Johnson, Samuel. *A Dictionary of the English Language.* London: W. Strahan, 1755.

Latour, Bruno. *We Have Never Been Modern.* 1991. Trans. Catherine Porter. Cambridge, MA: Harvard UP, 1993.

Laycock, Thomas. *A Treatise on the Nervous Diseases of Women.* London: Longman, Orme, Brown, Green & Longmans, 1840.

Law, C.M. 'The growth of urban population in England and Wales, 1801–1911.' *Transactions of the Institute of British Geographers* 41 (1967): 125–43.

Le Queux, William. *The Count's Chauffeur*. 1907. London: The Popular Library, n.d.

Lewes, George Henry. *The Physiology of Common Life*. 2 Vols. Edinburgh: Blackwood, 1859–60.

Martensen, Robert. 'When the brain came out of the skull: Thomas Willis, anatomical technique and the formation of the "cerebral body" in seventeenth century England.' *A Short History of Neurology: The British Contribution 1660–1910*. Ed. F. Clifford Rose. Oxford: Butterworth-Heinemann, 1999. 19–35.

McKendrick, John Gray. *A Review of Recent Researches on the Physiology of the Nervous System*. London, 1874.

Munro, H. H. ['Saki']. *The Chronicles of Clovis*. London: Penguin, 1948.

Murray, James, ed. *A New English Dictionary on Historical Principles*. 14 Vols. Oxford: Clarendon, 1888–1928.

Paré, Ambroise. *The Works of Ambroise Parey*. 1575. Trans. Thomas Johnson. London, 1691.

Payson Call, Annie. *Nerves and Common Sense*. 1909. Boston: Little, Brown & Co., 1912.

Philips, Edward. *The New World of English Words: Or, a General Dictionary*. London: E. Tyler, 1658.

Phormoid. 'Creates Flesh on Thin Bodies: Wonderful Powers Attributed to New Nerve Food.' Advertisement. *Pall Mall Gazette* 16002 (21 August 1916): 9.

Prochaska, George. *A Dissertation on the Functions of the Nervous System*. 1784. Trans. Thomas Laycock. London, The Sydenham Society, 1851.

Rabinbach, Anson. *The Human Motor: Energy, Fatigue, and the Origins of Modernity*. 1990. Berkeley: California UP, 1992.

Rapport, Richard. *Nerve Endings: The Discovery of the Synapse*. London: Norton, 2005.

Richardson, Charles. *A New Dictionary of the English Language*. London: Pickering, 1839.

Rousseau, George. *Nervous Acts: Essays on Literature, Culture and Sensibility*. Basingstoke: Palgrave, 2004.

Shepherd, Gordon. *Foundations of the Neuron Doctrine*. Oxford: Oxford UP, 1991.

Sheridan, Thomas. *A General Dictionary of the English Language*. London: J Dodsley, 1780.

Sherrington, Charles. *The Integrative Action of the Nervous System*. London: Constable, 1906.

Sinclair, May. 'The Novels of Dorothy Richardson.' April 1918. *The Gender of Modernism*. Ed. Bonnie Kime Scott. Bloomington: Indiana UP, 1990. 442–8.

Smart, B.H. *Walker's Pronouncing Dictionary of the English Language*. 3rd ed. London: Brown & Co., 1849.

Spencer, Herbert. *The Principles of Psychology*. 2 Vols. London: Williams & Norgate, 1870.

Spurzheim, Johann Gaspar. *The Anatomy of the Brain, with a General View of the Nervous System*. Trans. R. Willis. London: S. Highley, 1826.

St. Mars, F. 'Husband and Co.' *Pearson's Magazine* 21.181 (January 1911): 21–31.

Todd, Robert, and William Bowman. *The Physiological Anatomy and Physiology of Man*. 2 Vols. London: J.W. Parker, 1845–56.

Trotter, Thomas. *A View of the Nervous Temperament*. 3rd ed. Newcastle: Edward Walker, 1812.

Unzer, John Augustus. *Principles of a Physiology of the Proper Animal Nature of Animal Organisms*. 1777. Trans. Thomas Laycock. London: The Sydenham Society, 1851.

Verity, Robert. *Changes Produced in the Nervous System by Civilization, Considered According to the Evidence of Physiology and the Philosophy of History*. London: S. Highley, 1839.

Walker, John. *A Dictionary of the English Language*. London: T. Becket, 1775.

——. *A Critical Pronouncing Dictionary and Expositor of the English Language*. London, 1791.

——. *A Critical Pronouncing Dictionary and Expositor of the English Language*. 21st ed. London: Longman, Hurst, Rees, Orme & Brown, 1819.

Waller, Augustus. *The Nutrition and Reparation of Nerves*. London: Read & Co., 1861.

Webster, Noah. *A Dictionary of the English Language*. 1828. London; Black, Young & Young, 1832.

Wells, H.G. *The First Men in the Moon*. 1901. *The First Men in the Moon, The World Set Free and Short Stories*. London: Odhams, 1934. 5–163.

——. 'The Flowering of the Strange Orchid.' *The Complete Short Stories of H.G. Wells*. Ed. John Hammond. London: Phoenix, 2000. 9–15.

——. Manuscript for *The Time Traveller's Story*. H.G. Wells Collection, Rare Books and Manuscripts Library, University of Illinois at Urbana-Champaign. TT-001.

——. 'The New Accelerator.' *The Complete Short Stories of H.G. Wells*. 487–97.

——. 'The Sad Story of a Dramatic Critic.' 1895. *The Complete Short Stories of H.G. Wells*. 243–9.

——. 'A Story of the Days to Come.' 1899. *The Complete Short Stories of H.G. Wells*. 332–98.

——. *The War of the Worlds*. London: Penguin, 2005.

Whyte, Robert. *Observations on the Nature, Causes and Cure of those Disorders which have been commonly called Nervous, Hypochondriac, or Hysteric: To which are prefixed some Remarks on the Sympathy of the Nerves*. 2nd ed. Edinburgh: J Balfour, 1765.

Willis, Thomas. 'The Anatomy of the Brain.' 1664. *Five Treatises*. London, 1681.

Worcester, Joseph. *A Dictionary of the English Language*. London: Sampson, Low, Son & Co., 1859.

1
Beyond the Brain: Sceptical and Satirical Responses to Gall's Organology

Michael K. House

In the final decade of the eighteenth century, the Viennese physician Franz Joseph Gall (1758–1828) introduced a new natural-scientific theory that set out to explain the functions of the mind on a material basis. Gall's proposed physical account of the mind marked a radical methodological and theoretical reappraisal of the traditional dualism between mind and body. Earlier approaches, even those that sought a material contact-point between the body and the soul, began by positing an independent soul and then turned to the question of how this immaterial soul interacted with the material body. In *The Passions of the Soul*, for instance, Descartes proposed that the pineal gland served as the passive "seat of the soul" (352–3), which operated as a node that allowed the mind to both control cerebral and bodily activity and acquire sensory information. Gall's approach, on the other hand, bracketed the non-empirical substrate of the soul. His new psychophysiological science, which he termed both the 'doctrine of the skull' (*Schädellehre*) and the 'organology', argued that the diverse human faculties and inclinations are localizable in the discrete organs of the brain. Through empirical investigation he accounted for 27 such organs that corresponded to faculties as diverse as love of offspring, sense of pride, sense of colour and sound, poetical talent and the inclination to commit murder and thievery.[1] Based solely on the size of cerebral organs, reflected by the elevations or depressions on the skull, he claimed to be able to determine the strength of an individual's various faculties and inclinations. Thus, in opposition to the Cartesian dualist split, Gall provided a materialist-monistic account of mental activity that did not require a speculative leap into an immaterial realm.

Around 1800, materialism had a specific meaning, evoking an Enlightenment notion of a mechanistic structure in which an object is governed by the law of causality and, as such, lacks freedom. Thus, Gall's reduction of the human mind to grey matter was fated to generate great controversy. In December 1801, Emperor Franz II issued a decree banning Gall from publication and holding public lectures in Vienna. Combined with fears that people were too enthusiastic about Gall's work, the central reason for the decree was the potential that his theory could "lead to materialism and thereby go against the 'first principles of morality and religion'" (see van Wyhe 25). As Gall was unable to continue his work in Vienna he embarked, in March 1805, on an ambitious and, by all accounts, very well-attended two-year lecture tour throughout Germany and Europe.[2] The accusations of materialism, determinism and fatalism that he had encountered in Vienna echoed through the rest of Europe. Gall's materialistic-monistic account of the functions of the human brain threw the subject into a mechanistically determined natural order and in doing so threatened the possibility of free subjective self-determination at a time when Early German Romantic and Idealistic conceptions of the aesthetic, spontaneous individual sought to demonstrate the subject's independence from the causality of the mechanism of nature.

This chapter addresses two main strands of early Gall reception in Germany in non-scientific domains. First, within a philosophical context, it looks at the sceptical critiques of Gall forwarded by Gottlob Ernst Schulze (1761–1833) and G.W.F. Hegel (1770–1831). Schulze's critique relies on a notion of spiritual activity (*Geistesthätigkeit*), a term that implied the non-material basis of cognitive activity, and raises two central questions: first, what the material organization of the specialized organs of the brain can possibly express about spiritual activity; and second, how it is possible to account for development in human subjects if the various abilities and inclinations are considered innate. Hegel, who dedicates a chapter of the *Phenomenologie des Geistes* to a discussion of Gall's *Schädellehre*, goes still further, arguing that the skull and organs of the brain are fixed, contingent and naturally determined and, thus, can never reflect the free activity of subjective spirit. Furthermore, he argues that the reduction of the mind to an expression of bone (i.e. the skull) could only express the individual as a 'dead thing'. In Clemens Brentano (1778–1842) and Joseph Görres' (1776–1848) text, 'The Wondrous Story of BOGS, the Clockmaker', we see these philosophical critiques make their way into broader social satires. Through embedding a parodic form of Gall's theory in a society founded on a materialistic-mechanistic notion of the individual, they explore the dangerous social consequence of reducing the individual to a knowable, verifiable entity. Those elements of human nature resistant to the materialistic-mechanistically ordered society – which correspond to the production of fantastic images and aesthetic, predominately musical, inclinations – become a counterpoint to Gall's mechanistic account of consciousness. In all of these sceptical and satirical responses to Gall's organology, we see the drive to preserve the free self-determining integrity of the human subject through emphasizing the individual's non-reducible, non-objectifiable, non-static, non-graspable spiritual activity.

I

Gall was quite aware of the controversy his materialist conception of the mind would generate. In late 1798, a letter outlining his project appeared in Wieland's popular journal *Der neue Teutsche Merkur*. The letter, addressed to a Viennese censorship official, Baron Joseph von Retzer, served as both an exposition of his system and a rejoinder to some early attacks. In the letter, he focuses on two central issues raised by his critics: first, whether his system excludes a notion of an immaterial mind or soul; and second, whether individuals are slaves to, in the sense that they were determined by, the innate, material structure of their cerebral organs and, thus, lack free will. In his response, Gall does not tone down his rhetoric of material determination; instead, he demonstrates how it is possible to explain notions of mental activity and the educatability of man from a material basis.

Before addressing Gall's response to his early critics, it is important to see how he frames his project. Early in the letter he presents the scope and function of organology as follows:

> my purpose is to determine the functions of the brain in general, and those of its different parts in particular; to show that it is possible to determine different abilities and inclinations by the elevations and depressions on the head, and to present in a clear way the most important truths and consequences to medicine, morality, education, legislation, and so on, and generally, to the science of human nature. (312)

Here we see that understanding the structures of the brain and with them the various cognitive functions stands at the foundation of his science but does not mark the limit of its application. He strategically situates organology as a foundational science that contains the potential for universal application. This is particularly evident in his dismissive treatment of Kantian terminology, which had come to dominate philosophical and scientific discourse in the last decades of the eighteenth century. He sarcastically writes:

> You must not blame me, that I have not used Kantian language. I have not yet come so far in my research to discover a special organ for acuteness [*Scharfsinn*] and profundity [*Tiefsinn*], for the principle of the ability of presentation [*Vorstellungsvermögen*] for the various types of ability of judgments [*Urtheilsvermögen*] etc (313–14).

As we read on, it becomes clear that Gall never seeks to provide empirical verification of Kant's transcendental subject, rather, for Gall, "the path of experience" (332) is the basis of his scientific pursuit. Even his use of the term 'experience' places him in opposition to Kant's transcendental project, in particular, Kant's account of the *a priori* foundation for cognition. Gall writes: "I have not sufficiently appreciated the *a priori*, that is to say, the philosophy which is to be founded upon the *a priori*" (332). We can only read this as a subversion of Kant's premise from the introduction to his *Critique of Pure Reason*: "even though all our cognition starts *with* experience, that does not mean at all that it arises *from* experience" (Kant A1; B1). For Gall, a complete account of the mind can occur from within experience, from within the empirically visible structures of the brain.

Gall's polemic against the philosophical approach to the question of human cognition reveals both a methodological and a theoretical domain within which he situates his account of the subject. Gall always portrays himself, as John van Wyhe points out, as "a follower of Nature, whose findings were uniquely determined by that irrefragable source of truth" (21).[3] This involved bracketing metaphysical questions of the soul, and, as Gall's collaborator, J.G. Spurzheim remarked, "we never endeavour to explain final causes; but have always declared, and every where do declare, that we make no inquiry into the nature of the soul, nor into that of the body: we are led only by experiment" (492).[4] Gall's invocation of a methodological constraint serves both to orient his science and to withdraw it from questions of speculative metaphysics. Specifically, he wishes to subvert the question of 'spiritual nature', which he sees his critics formulating as follows: "If the functions of the soul [*Seelenverrichtungen*] are established by corporeal instruments

[*Werkzeuge*], or organs, are not the spiritual nature and the immortality of the soul called into question?" (Gall 319). And in his answer we see the dual strategy of justifying his empirical approach and bracketing all that remains outside of the empirical: "The student of nature seeks to obtain only the laws of the material world, and presupposes that no natural truth can be in contradiction with any established truth.... He merely perceives and teaches that in this life the mind is chained to a corporeal organization" (319). Here we see that his invocation of a methodological limitation on his project corresponds to a theoretical assertion of mental limitation. More specifically, the mind becomes chained to its corporeal reality, and, thus, the material form of the brain (and skull) establishes a limit on the potential activity of the mind.

He was also aware that the limitation of the mind to its cerebral 'instruments' spurred concern that he was forwarding forms of materialism, natural determinism and fatalism. For his opponents the questions of freedom, of distinguishing between good and evil, and of human agency were all at stake. In the letter, Gall responds to the concern in terms of human educatability. Gall, as Claudio Pogliano presents it, "recalls how the agent remained distinct from the material means, and how action should not be identified with the faculty to act" (152). At first glace, Gall's response to his opponents verifies Pogliano's remarks:

> Those who wish to convince themselves that our qualities [*Eigenschaften*] are not innate, attribute them to education. But have we not alike acted passively, whether we have been formed by our innate dispositions, or by education? With this objection, they mistake the ideas of faculties, inclinations, and simple disposition, for the mode of action [*Handlungsweise*] itself. (315)

However, Gall's argument does not attribute a notion of agency to a non-material source independent of the brain. What are traditionally treated as non-natural inclinations and senses belong, with Gall, to the material constitution of the subject:

> [M]an possesses, besides the animal qualities, the faculty of speech, and the most extensive ability for education [*Erziehungsfähigkeit*] two inexhaustible sources of knowledge and action [*Bewegungsgrund*]. He has the sense [*Sinn*] of truth and error, of right and wrong, for the presentation of an independent being; the past and the future guide his action; he is endowed with moral feeling, with conscience, etc. (315)

Thus, the inclinations towards morality are themselves a product of the cerebral organs, and even though he speaks of internal conflict, even of self-denial (*Selbstverläugnung*), this refers to a material conflict between organs, that is, the power of one organ over another. It is the space for self-denial or self-negation – here understood as a negation of material pre-dispositions – that becomes central for early readers of Gall. For Gall, however, the space for human agency, freedom, even arbitrary action (*Willkühr*) still remains within the material constitution of the discrete organs.

II

In 1807, Gottlob Ernst Schulze reported on Gall's lecture tour in a text titled 'Concerning Gall's Discovery of the Organs of the Brain' for the journal *Chronik des neunzehnten Jahrhunderts*. Schulze was best known for his publication of two earlier texts: *Aenesidemus*, a sceptical attack on Kant's first *Critique*, which appeared anonymously in 1792, and a continuation of his sceptical project in his 1801 text, *Kritik der theoretischen Philosophie*. In both texts, he demonstrates the limitation of philosophy heretofore to establish the connection between our subjective representations and the way external reality is 'in-itself'.

In his reading of Gall's organology, this translates into a discrepancy between the form of an organ and the function this form supposedly designates. While he grants that much of what Gall teaches about the form of the cerebral organs appears to have empirical validity, the organological model cannot hope to explain the function or employment of these organs. Gall can, according to Schulze, explain the physical layout of the organs and their relation to cranial formation, but their function – which, he understood as a 'spiritual activity' – and the human capacity to acquire new talents (i.e. change over time) eludes him. Schulze writes:

> [T]he conclusion [that the interior structure of the organs is reflected in the external surface of the skull] leads only to the knowledge of mere constructions [*Anlagen*] and aptitudes [*Befähigungen*] for given spiritual activities [*Geistesthätigkeit*], but not to the insight of acquired skills [*erworbenen Fertigkeiten*] and character of a man. (1129)

Schulze argues that a fundamental distinction must be made: "the particular construction of an organ ... must be distinguished, namely from the education/development [*Ausbildung*] and the activity [*Wirksamkeit*] of the organs that takes place in life" (1129). The conclusion of this portion of his argument emphasizes the importance of education (*Erziehung*), the counterpoint to material determination forwarded by Gall's theory. Schulze's critical presentation of Gall's work limits what the distinct organs of the brain can explain about the individual through developing a notion of activity, of activation, that does not have its source, its 'cause', in the natural structures of the brain. Moreover, he presents examples in which the predominance of a particular organ still requires education for its actualization or negation. For example, an individual with an enlarged organ for the sense of sound requires *Ausbildung* before becoming a talented musician, and conversely, the prevalence of the so-called 'thievery-organ' (*Diebsorgan*) or 'murder-organ' (*Mordorgan*) can be reduced or negated through *Ausbildung* to avert a life of thievery or murder (1129–30).

Before departing into a questionable critique of the Eurocentric limitation of Gall's empirical study,[5] Schulze attempts to both restrict the work of Gall within the realm of an inductive empiricism and open an indeterminate space that he has already demarcated as that of independent spiritual activity. Schulze sees the former point as implicit in Gall's work: "it is, even according to Gall, a misunderstanding of the doctrine of the organs, if one believes, that one can read through the medium of the exterior surface of the skull of a man what transpires in his interior" (1134). Gall's "advocacy of the possibility of character analyses based on head-reading", which, as van Wyhe points out, "attracted the most attention" (23), contradicts Schulze's argument. Here, and as we will see with Hegel, Brentano and Görres, it is the very possibility of 'reading heads' to determine 'what transpires in the human interior' that they call into question. They rigorously interrogate both what, if anything, the organs and skull reveal about an individual, and also, if Gall's theory can perform the task it sets out for itself, what this implies for the subject.

In the *Phenomenology of Spirit*, published at the end of Gall's tour in 1807, Hegel takes up this very line of questioning. His polemical, sceptical account of Gall's theory, found in a chapter titled 'Observation of the Relation of Self-consciousness to its Immediate Actuality. Physiognomy and *Schädellehre*', not only argues, as does Schulze, for the limitation of Gall's system, but goes further, to demonstrate its complete inability to express anything regarding what he calls the reality of the individual. Concerned more with his polemic against empirical-naturalist accounts of self-consciousness and the individual than with presenting an accurate account of Gall's system, Hegel ignores some fundamental points about the causal relation between the interior brain and its exterior expression in the skull and focuses primarily on the idea that the skull expresses the reality of the individual.[6] If this were true, Hegel argues, the individual spirit would be reduced to the contingency of nature and not any element of nature, but the dead structures of the inorganic. In the end, Gall's *Schädellehre* represents the unforgivable enterprise of turning the "reality of Spirit itself into a Thing or, expressing it the other way round, gives to lifeless being the significance of Spirit" (Hegel 208).

Hegel expresses his central argument in highly polemical and sarcastic terms: "the individual can be something else than he is by inner disposition, and still more than what he is as a bone" (204). That the individual is more than bone, more than the external shape of his skull, belongs to his fundamental claim that the skull does not express the free activity of the individual, but only the intrinsic, given the beginning point of the dialectic of self-consciousness:

> The individual exists in and for himself: he is *for himself* or is a free activity; but he has also an *intrinsic* being or has an *original* determinate being of his own.... In his own self, therefore, there emerges the antithesis, this duality of being the movement of consciousness, and the fixed being of an appearing actuality, an actuality which in the individual is immediately his *own*. (185)

Thus, at best, Gall can account for the original determination of a subject's being as a static, determinate being of actuality, but by such an account cannot derive an explanation for the free, self-determined 'movement of consciousness' that he sees as fundamental in determining the individual.

The idealist notion of individual activity (*Selbsttätigkeit*) forms the ground for the individual: "the true being [*das wahre Sein*] of a man [is] deed" (Hegel 193). As a 'thing' the skull is a contingent natural construction that neither encapsulates nor limits the activity of spirit (i.e. the being of man as deed). Thus, for Hegel, the contingency of the material construction of the skull can neither explain an action nor has even symbolic value, as a material manifestation expressing an immaterial spirit: "We neither commit theft, murder, etc. with the skull bone...nor has this *immediate* being the value even of a *sign*" (200–1). The skull-bone, Hegel states later, "qua sign" is "indifferent to the content it is supposed to denote" (207). The skull cannot stand in as a meaningful, external expression of the interior reality of consciousness.

Thus, 'reading heads' is a meaningless activity. Hegel, however, acknowledges that even as a meaningless activity, it carries a consequence: "When, therefore, a man is told, 'You (your inner being) are this kind of person because your skull-bone is constituted in such and such a way,' this means nothing else than, 'I regard a bone as *your* reality'" (205). Allowing such statements reduces the reality of the individual to the same status as a dead thing.

In this context, Hegel also defines the reality of an individual with regard to its materiality. In doing so, he goes so far as to grant that the material determination of the subject does play a part in the constitution of consciousness, but only insofar as the spiritual activity negates the material basis, and becomes an independent free self-relation: "What merely *is*, without any spiritual activity is, for consciousness, a Thing, and, far from being the essence of consciousness, is rather its opposite; and consciousness is only *actual* to itself through the negation and abolition of such a being" (205). Thus, in the course of his critique of Gall's *Schädellehre*, Hegel develops a notion of individual consciousness as the negation of existence as a mere thing. Inversely understood in terms of the method of knowing another individual, the science of man cannot then begin from the external being of material structures (i.e. the structure of the skull) and attempt to attain this "true being" of an individual as thing.

III

While Hegel and Schulze's critiques of Gall emphasize the limitation of an empirical grounding of a subject's mind and even go so far as to argue that a material explanation cannot adequately account for the 'spiritual activity' of the subject, two early Romantic authors, through a literary, satirical employment of Gall's method, explore the dangerous social consequences of the claim to verifiable empirical knowledge of the individual. In the concluding scene of Clemens Brentano and Joseph Görres' 1807 work *The Wondrous Story of BOGS, the Clockmaker*,[7] the authors combine Gall's theory on the organization of the discrete organs of the brain with Phillip Bozzini's newly invented light conductor (*Lichtleiter*), a prototype for the endoscope.[8] The scene takes the form of a 'Visum Repertum' or doctor's report on the condition or 'wellbeing' of the protagonist's brain. Through the satirical employment of contemporary medical discourse, Brentano and Görres undermine the assumptions of natural science that grant privileged access to the human mind from an empirical-materialistic basis.

As the doctors, following their Gallean map, burrow into the cerebral activity of BOGS' brain, they do not encounter scientifically readable images, but rather fantastic forms and figures (also described as insane, erratic and incomprehensible) resistant to their means of exploration and explanation. The purpose of this investigation is to determine whether BOGS should be accepted into civil society, a stipulation of which is that he eradicates all of his tendencies towards the aesthetic, the musical and the 'fantastic'. Thus, the incomprehensible forms and figures that undermine the doctors' pursuit are the same forms and figures that make him 'incompatible' with the social. The authors thereby conflate a critique of materialistic models for the human intellect with a critique of the conformism of social institutions.

The text begins with a public announcement: a protocol, written in a satirical *Amtstil* or bureaucratic style, outlining the structure and new order of society, or the *Schützengesellschaft*.⁹ Within the protocol, the authors set up a dichotomy that places questionable Enlightenment social conceptions derived from a combination of mechanistic materialism and a Leibnizian notion of a fixed 'pre-established harmony' on one side, and the newly emerging Romantic conception of spontaneous, creative, aesthetic individual on the other. They align their notion of the human with the latter, giving a greater level of authenticity to the Romantic conception of the individual. In the protocol BOGS learns that the *Schützengesellschaft* is calling for the "elimination of all remaining traces of humanity [*Menschlichkeit*]" and that "world and life" be replaced with "country and state" (435). To this end, the protocol demands that all "prophets, wise-men, philosophers, enthusiasts or visionaries, poets, musicians, painters and artists", all of those in the business of producing "useless fantasies . . . [and] fabulous stories" (435) report to the local authorities with a written "confession [*Selbstbekenntnis*] about their character and principles [*Grundsätze*]" (436). The amalgamation of a drive to know the individual (or for the individual to make himself known) and the systematic destruction of all elements that do not correspond to the system's own design form a critique of social and as the text progresses, scientific institutions that desire only to know what they presuppose.

Amenable to the demands of the protocol, the protagonist produces a confession and it is here that the materialistic-mechanistic conception of the individual first emerges. As a child, BOGS recounts that he would exclaim: "do not let me fall under any gear, so that I myself may become a good gear, or a healthy spoke" (436). His acquiescence finds expression in his occupation. He becomes a clockmaker in order to be a good citizen in a world that he views as taking "the form of a well-set clock", anticipating the *Schützengesellschaft*'s threat that "those whose chains and gears are not already attached [to the social system], will be chained and made into gears" (436). The metaphor of the clock appeals to numerous Enlightenment mechanistic conceptions of bodies in the natural world. In the sixth *Meditation*, Descartes speaks of the laws of nature as "a clock made of wheels and counter-weights" and regards "man's body as a kind of mechanism that... if no mind existed in it, the man's body would still exhibit all the same motions that are in it now except those motions that proceed from a command of will or, consequently, from the mind" (55).[10] The authors of BOGS, however, undo Descartes' exclusion of the mind and will from the mechanistic model through BOGS' assertion that, "our soul is metal" (438).[11] Furthermore, the mechanistic functionality becomes the basis for understanding the human soul (both as metal and as spoke) with a social model (society as clock) in which, as Caroline Welsh argues, the protagonist understands himself and is understood by the social institution as "a human automaton" (253). Here, Hegel's critique of a materialistic notion of spirit as dead thing (soul as metal) has the consequence of lost autonomy and social complicity.

The narrative tension arises when it turns out that BOGS admits that he does not entirely conform to the mechanistic model to which he aspires. He has an uncontrollable "madness for music [*Tollheit über Musik*]" (441), an unacceptable character trait in the *Schützengesellschaft*. They instruct BOGS to attend a concert and provide a written report, yet another test to determine his compatibility. As the report's prose disintegrates into a chaos of fantastic and nonsensical imagery, the reader begins to sense that BOGS' chances for inclusion are diminishing. However, his resistance to the well-ordered structures of the society makes him an object of curiosity and with the dual purpose of the "advancement of science" and "examination" BOGS is instructed to give himself over to the "*Consilio medico*", after which "more will be decided" (454).

The examination begins with the discovery of a contusion on the top of his head. The contusion, the doctors explain, is the result of BOGS' "unfortunate fall" at birth from the absolute or ideal, in which he "hit the real" (455) head first. In determining that the contusion explains the damaged "integrity of his ability to understand [*Verstandeskräfte*]" (455) and that this defect is *inborn* establishes it as the first allusion to the materiality of cognitive abilities forwarded in Gall's *Schädellehre*. This "anomaly from a normal condition" – verified by his "abnormal polarity" (455), a reference to Mesmer's theory of animal magnetism – compels the doctors to look closer at the candidate's head, and they discover yet another divergent structure that will prove an obstacle for their investigatory apparatus. BOGS is Janus-faced, having a second distinct face under his hair with a unique, oppositional disposition. From the formation of his skull (*Schädelbildung*), an explicit appeal to Gall's *Schädellehre*, they find him to have a predominant "sense of loftiness, sense of profundity, arrogance, humility, gravitas, loftiness, murderousness, peacefulness, thievery, and sense of catching thieves" (457). Corresponding to the newly discovered 'second' or 'anti-BOGS', the strength of one faculty, marked by an "elevation" in his skull, is annulled by a "depression", and their ability to determine the "authentic nature and composition of the subject" (457) is further frustrated.

With the failure of the external, superficial examination to produce definitive results, the investigation moves to the Gallean source of these structures, the interior of the skull. To prepare BOGS for the organological investigation they first hypnotize him, or more accurately mesmerize him, by lulling him into a state of 'magnetic sleep' or unconsciousness. After they hypnotize BOGS, he does not become sedate; rather, the unconscious state triggers a series of violent convulsions. More precisely, BOGS begins to "fantasize in paroxysms" (458). They describe the paroxysmal fantasies as "a stream of nonsense and foolish gibberish [flowing] out of both mouths" that certainly could "not be believed... had not the noble members of the *Schützengesellschaft* heard it themselves" (458). The authors employ this figure of convulsion to demonstrate the resistance of the subject to the operations of inquiry. Through a convulsive shaking, BOGS becomes theoretically uncontainable, and it is this figure that undermines the quest both to localize BOGS within the social system, and in terms of Gallean organology, to localize the origin of his cognitions.

After BOGS eventually settles, they insert Bozzini's *Lichtleiter*. Following the "olfactory nerves" and entering the frontal brain cavity, which is illuminated "perfectly and brightly and clearly", the light conductor reveals "things of wonder [*Wunderdinge*]" (458). What their endoscopic gaze reveals cannot find expression in cold, scientific prose; instead, they invoke the Romantic language of aesthetic experience and describe their experience in terms of an "ineffably sweet and sublime feeling" (458). The meeting of these two discourses becomes a scene of violence. What appears through the lens of the *Lichtleiter* is not grey matter, but images of owls and bats never before disturbed from their serene "darkness", which, exposed to the light, "flock together in fear and topple forward [*hervorstürzen*]" (459). These images, in turn, are of "great horror [*Schrecken*] to the viewers", the authors write, "who had not expected to encounter such obscene and repulsive vermin [*Geschmeiß*] in man's noble organ" (459). The resolution of this horror and disappointment occurs at the hand of Bozzini's light conductor, the light of which finally "burns them away", "effacing" them (459). The text repeats the violence against illegible figures later as they inspect the "hearing nerves" of "the fourth cavity in the brain" (459), a predominance of which, Gall argues, corresponds to musical abilities. The doctors find anthropomorphized tones or musical notes that "run nonsensically…around the walls" of the skull and "do not lend themselves in any way to comprehension, and all amicable attempts at representation were fruitless" (459–60). The solution is the application of glue, described as a "means of coercion [*Zwangsmittel*]" (460) that serves to stick the wings of these musical notes and tones. Here the rapid, agitated movement of these 'nonsensical' musical entities is itself arrested, which has the outcome of death: it is reported that these musical figures "are brought through the deprivation of water from life to death" (460). The 'investigation' by this apparatus of the *Schützengesellschaft* is not a mere 'observation', but an act of control, as they continue removing, sedating and eliminating those elements in BOGS' brain that are at odds with both their own scientific (pre)conception of the human, which is likewise the society's conception of the citizen. And rather than determining whether BOGS is suitable for the society, the acts of annihilation and subversion resulting from their invasive procedures become tantamount to an attempted lobotomy of BOGS.

The single discovery that does not cause the doctors to employ their "means of coercion" is of "hundreds of thousand of microscopic clocks" (459), lining the walls of the frontal portion of BOGS' skull. This, of course, is also the single aspect of BOGS' brain that the inquirers leave untouched, as it both confirms their preconception and aligns perfectly with the mechanistic social conception presented in the public protocol. This is where they observe BOGS' "subtlety of human understanding" (460), an accurate application of Gall's model, which places the uniquely human faculties in the frontal portion of the skull. The reappearance of the metaphor of the clock establishes a direct correspondence between the structure (or anatomy) of the organs of thought and the structure of a 'well-ordered' society. This doubling of metaphors corresponds to a notion of a society striving towards a totalizing self-identity, that is, a remainderless identity between the social system and its components/citizens. The authors re-enforce this idea through the appearance of a 'little man [*Männchen*]' who is assigned the task of continuously winding and cleaning the clocks (i.e. maintaining BOGS' internal harmony). Described as "the little man, who we see when we look another man in the eyes", he "appeared to have control [over the network of clock]" (459). The man operates as the singular, divine cause in an occasionalist model, maintaining order and harmony within BOGS' brain, but as a reflection in his eyes, he is also the image of the doctors, or the *Schützengesellschaft* itself, and thus, the man becomes a mechanism of control, directly implanted in the mind's organs.

Although the text establishes a proto-Foucauldian argument for the scientific model's dual function as a mechanism of inquiry and control, it also demonstrates the subject's resistance to such mechanisms. The promised interior perspective provided by their technological apparatus fails, as do the scientific models they employ. The *Lichtleiter* cannot illuminate the inner depths of BOGS' brain and what they find in the "fourth [and last] brain cavity" becomes "obscure and uncertain" (462). This portion of the brain, which they, in a state of complete uncertainty, speculate is the site of potential "nigromancy", contains, insofar as the "darkness allowed them to see anything at all," what they are only able to describe as "strange characters" (462). These uncanny characters are without certain qualities, and the authors begin designating this space an abyss for understanding. The rhetoric of the limit of understanding ultimately becomes intermingled with the rhetoric of the "limits of caution" (492), both of which one of the doctors oversteps, when, in yet another strange turn, he decides to enter the brain himself, torch in hand. The improbable quest of the doctor after "the truth" (462), likened to Orpheus after his Eurydice, ends in disaster. Fireworks, explosions, a great chaos of destruction, all burst forth out of this abyss of human understanding. Again the notion of the paroxysm returns, and this time the "paroxysm reaches its highest apex, and the movement becomes so strong that the centrifugal force of the swaying" (463) sucks the doctor into the arms of an interior embodiment of cholera, one of the four temperaments, and the doctor becomes a "regretful sacrifice of his own intellectual curiosity" (464). This scene of doubt, darkness and abyss marks the other side of the human intellect and by losing the doctor, the agent of their medical-scientific inquiry, in this space the authors mark a rupture in our ability to attain fully the underlying source, or foundation of human intellect through materialistic-scientific models.

Brentano and Görres share in Schulze and Hegel's postulation that if Gall's materialist reduction of human mind is correct, then the mind is constrained and determined by the causal laws of the mechanism of nature and no longer retains the power of free self-determination. Their interventions are also alike insofar as they attempt to limit Gall's project and demonstrate an element of the subjective mind that eludes explanation on a material basis. For Schulze and Hegel, 'spiritual activity' designates this element, an ambiguous term that casts the freely self-determined agency of subjective consciousness to an ambiguous realm outside of the materially constituted cerebral organ(s). While their goal is not to completely undermine the function of the brain as an intermediary between the spiritual and the material, their attack focuses on the question of what these cerebral organs can possibly express. Brentano and Görres' critique does not simply point to the inadequacy of the organs to express the reality of the mind, but also complicates any theory that claims to compartmentalize and organize the mind according to a rationalistic-materialistic system by pointing to a dimension of the human mind that actively resists and even undermines such determinations. They accomplish this in their fictional exploration of the brain by including the cryptic figures or 'strange characters' that spontaneously emerge from an inner depth of the mind to frustrate the apparatus of scientific inquiry. These figures allude to a space within the subject capable of producing such figures but stop short of providing an explanation of their source. This indeterminate space acts as a safeguard against the text's conclusion that fixing the subject within a materialistic-scientific project is tantamount to divesting it of its integrity, autonomy and free self-determination. Thus, in order to preserve the subjective freedom they create a source of imaginative creation that not merely resists mechanistic, rationalistic and systematic structures, but actively disrupts them.

In short, Brentano and Görres create a notion of the unconscious. Specifically, the inner depth of the human interior becomes a repository for inclinations and desires that exceed or even run contrary to both models of the rational individual within a well-ordered society. This recalls what L.L. Whyte calls "the fundamental agent, the ultimate driving force, behind the discovery of the unconscious", namely, "that element of surplus vitality or refusal to be content with life as it is, which had the power to force self-conscious man to transcend his image of himself, to become richer as a person by recognizing the limitations of his current idea of himself" (69). The objection held by Schulze, Hegel, Brentano and Görres alike is that a comprehensive systematic physical account that claims to capture the mind as it is imposes a limit on the potentiality of the mind and thereby threatens not just its surplus vitality but any vitality. The reduction of the mind to something expressible in the lifeless matter of the bone (Hegel) or to the metallic order of a machine (Brentano and Görres) finds its counterpoint in the life of an uncontainable excess that does not allow itself to be reduced to any system.

This early appeal to the unconscious, understood as the unrestricted and uncontrolled site for irrational, nonsensical production, attempts to embed within the human psyche a moment that undermines the mechanistic and rationalistic framework for understanding the mind. Specifically, with the incomprehensible images and figures emerging from what can be described as the chaotic madness of the human interior the authors establish something inherent in the human mind that destabilizes a notion of the individual restricted to rational consciousness as well as social orders based on the limited understanding of the individual as rationally determined. This destabilization functions not so much to destroy or completely dismantle the notion of the rational individual, but rather it brings to light the inherent incomprehensible and divergent impulses that rational consciousness represses and civil society suppresses. Thus, in the effort to counter Gall's theory they create a dimension of the subjective mind that is both a systematic component of the mind and disruptive to any systematic understanding of the mind.

Notes

1. For a complete list and map of Gall's 27 organs of the brain, see Ackerknecht and Vallois (24–5).
2. For more details about the tour, see van Wyhe (28).
3. See also Clarke and Jacyna (228–30); Pogliano (152) and Oehler-Klein (114–28). Clarke and Jacyna cite Gall's early exposure to Herder, in particular his *Ideen zur Philosophie der Geschichte der Menschheit* (1784–91) as the source for his "dynamic and vitalistic" concept of nature, "whereby unity of structure and function was a basic principle of both the inorganic and organic world" (229).
4. Ackerknecht and Vallois cite Gall making the same point in an even later work: "I am not trying to explain the essence of any of the faculties (of bodies and souls), I limit myself to phenomena . . . I am not explaining causes, I indicate conditions" (qtd. 19).
5. Considering the dubious, racial tendencies of the history of phrenology, it is perhaps not surprising that here Schulze's critique attempts to account for the lack of cultural development of African "Buschmänner von Hottenttotten", which he views as a shortcoming of Gall's work.
6. For an account of the limitations of Hegel's presentation on Gall perhaps the best sources are Oehler-Klien and Drüe.

7. The actual title of the text is: 'Entweder wunderbare Geschichte von BOGS dem Uhrmacher, wie er zwar das menschliche Leben längst verlassen, nun aber doch, nach vielen musikalischen Leiden zu Wasser und zu Lande, in die bürgerliche Schützengesellschaft aufgenommen zu werden Hoffnung hat, oder die über die Ufer der Badischen Wochenschrift als Beilage ausgetretene KONZERT-ANZEIGE; Nebst des Herrn BOGS wohlgetroffenem Bildnisse und einem medizinischen Gutachten über dessen Gehirnzustand.'

8. See Bozinni.

9. Elisabeth Stopp points out that, at the time, the term that could be translated as a society of marksmen, a society of burglars, or even a society of bunglers and is taken from Jean Paul's satire *Siebenkäs*, in which it is used to denote an "archetype of civil conformism" (360).

10. The clock metaphor is also a central part of Leibniz's description of pre-established harmony, in which he asks his reader to imagine "two clocks in perfect agreement", and argues that the agreement is maintained because God initially creates them with such perfection and precision that they remain forever consistent (147–9).

11. Caroline Welsh accurately points out that this could be an allusion to Le Mettrie's notion of the materiality of the soul (257).

Works cited

Ackerknecht, Erwin H., and Henri V. Vallois. *Franz Joseph Gall, Inventor of Phrenology and His Collection*. Trans. Claire St. Léon. Madison: Dept. of History of Medicine, Univ. of Wisconsin Medical School, 1956.

Bozzini, Phillip. 'Lichtleiter, eine Erfindung zur Anschauung innerer Theile und Krankheiten nebst der Abbildung.' *Journal der practischen Arzneykunde und Wundarzneykunst* 24 (1806): 107–24.

Brentano, Clemens and Joseph Görres. 'Entweder wunderbare Geschichte von BOGS...' *Clemens Brentano. Werke in zwei Bände*. Ed. Friedhelm Kemp and Wolfgang Frühwald. Munich: Carl Hanser Verlag, 1972. Vol. 1.

Clarke, Edwin, and L. S. Jacyna. *Nineteenth-Century Origins of Neuroscientific Concepts*. Berkeley: California UP, 1987.

Descartes, Rene. *Meditations on First Philosophy*. Trans. Donald A. Cress. Indiana: Hackett, 1993.

Descartes, Rene. *The Passions of the Soul*. Trans. J. Cottingham, R. Stoothoff & D. Murdoch. *Selected Philosophical Writings*. Cambridge: Cambridge UP, 1988.

Drüe, Herman. *Psychologie aus dem Begriff. Hegels Persönlichkeitstheorie*. Berlin: de Gruyter, 1976.

Gall, Franz Joseph. 'Schreiben über seinen bereits geendigten Prodromus über die Verrichtungen des Gehirns der Menschen und der Thiere, an Herrn Jos. Fr. von Retzer.' *Der neue Teutsche Merkur* 3 (December 1798): 311–32.

Hegel, G. W. F. *Phenomenology of Spirit*. Trans. A.V. Miller. Oxford: Oxford UP, 1977.

Kant, Immanuel. *Critique of Pure Reason*. 1781. Trans. Ed. Werner, S. Pluhar & Patricia Kitcher. Indiana: Hackett, 1996.

Leibniz, Gottfried Wilhelm von. 'Postscript of a Letter to Basnage de Beauval.' *Philosophical Essays*. Ed. and trans. Robert Ariew and Daniel Garber. Indiana: Hackett, 1989.

Oehler-Klien, Sigrid. *Die Schädellehre Franz Joseph Galls in Literatur und Kritik des 19. Jahrhunderts*. Stuttgart: Gustav Fischer Verlag, 1990.

Pogliano, Claudio. 'Between Form and Function: A New Science of Man.' *The Enchanted Loom: Chapters in the History of Neuroscience*. Ed. Pietro Corsi. New York: Oxford UP, 1991.

Schulze, Gottlob Ernst. 'Ueber Galls Entdeckungen die Organe des Gehirns betreffend.' *Chronik des neunzehnten Jahrhunderts* 2 (1807): 1121–52.

Spurzheim. Johann Gaspar. *The Physiognomical System of Drs. Gall and Spurzheim; Founded on an Anatomical and Physiological Examination of the Nervous System in General, and of the Brain in Particular; and Indicating the Dispositions and Manifestations of the Mind*. London: Baldwin, Cradock, & Joy, 1815.

Stopp, Elisabeth. 'Kunstform der Tollheit.' *Clemens Brentano. Beiträge des Kolloquiums im Freien Deutschen Hochstift*. Ed. Detlev Lüders. Tübigen: Max Niemeyer Verlag, 1978.

van Wyhe, John. 'The Authority of Human Nature: The Schädellehre of Franz Joseph Gall.' *British Journal for the History of Science* 35 (March 2002): 17–42.

Wegner, Peter-Christian. *Franz Joseph Gall, 1758–1828: Studien Zu Leben, Werk Und Wirkung*. Hildesheim: Georg Olms Verlag, 1991.

Welsh, Caroline. *Hirnhöhlenpoetiken: Theorie Zur Wahrnehmung in Wissenschaft, Ästhetik Und Literatur Um 1800*. 1. Aufl. ed. Freiburg im Breisgau: Rombach, 2003.

Whyte, L.L. *The Unconscious before Freud*. London: Julian Friedmann, 1978.

2
Neurology and the Invention of Menstruation

Andrew Shail

> I do not see how it is possible to advance any speculation as to the *cause* of menstruation other than it can exist in a nervous mechanism
>
> Robert Lawson Tait, *Diseases of Women*, 1889, 322

Theories of menstruation provide a fundamental site for both the expression and the development of notions of sexed being. This chapter examines various neurological accounts of menstruation produced during the late eighteenth and nineteenth centuries to outline the origins of one of the core building blocks of the period of historical modernity: the notion that the female body is an essentially well-functioning entity. Between 1875 and 1886, a number of gynaecologists advanced the major components of our current understanding of the purpose of menstruation: it occurs at the opposite end of a monthly cycle from ovulation, and it accompanies the shedding of the epithelium (the tissue comprising the lining of most inner cavities) of the uterus, now known as the endometrium. Even this discovery, however, held that menstruation was pathological (albeit systemically so). As Robert Lawson Tait put it, summing up Wilhelm Löwenthal's findings published in German in 1884 and 1885, "menstrual bleeding is neither a physiological function nor an accompaniment of one, but it is a consequence made habitual by innumerable repetitions of a state of things artificially produced – viz., the non-impregnation and death of the egg; it has all the peculiarities and effects of other undoubtedly pathological hæmorrhages" (322). Where most animals, it was recognized, re-absorb the cells of a denuded uterine epithelium, the occurrence of uterine bleeding to 'flush' the epithelium out of the body was virtually unique to humans (Tait 332).

Nonetheless, although this discovery of normalcy incorporated the pervasive urge to find fault with the female body, the gynaecologists of the era were not conscribed solely by this will to pathologize. Bland Sutton and Arthur W Johnstone's studies, published separately in 1886, proved, Tait concluded, that "the main factor of the process [menstruation] is the preparation of this uterine mucous surface for the retention of the ovum, after its fertilization, by the denudation of its epithelium" (324). Even having established the role of menstrual bleeding in the removal of the uterine epithelium, gynaecologists could describe the part of this removal in the monthly cycle as *preparing* the uterus for the retention of an ovum. While, in the West, there has since been a widespread education in the 'knowledge' that menstruation occurs *after* ovulation and so therefore signals the *failure* to get pregnant, Sutton, Johnstone, Tait and their contemporaries could just as easily interpret their evidence as showing that menstruation occurs *before* ovulation and so enables the capacity to become pregnant. This chapter argues that a major impulse behind this potential to see the female body as essentially well-functioning was the emergence of the modern neurological notion of sex.

One reason why the 'discovery' of the purpose of menstruation occurred as late as the 1880s was that, for most nineteenth-century gynaecologists, the ovaries were the dominating factor in the female economy. By 1840, Thomas Laycock described a number of contemporaries who were sure of a ubiquitous "physiological and pathological action of the ovaria and testes on the system in general" (ix). Under the aegis of this conviction in the primacy of the ovaries, menstruation was linked to ovulation as its simple and immediate consequence.[1] Although ovulation was not discovered until 1827, when Karl von Baer first described the mammalian ovule, elements of the theory that the ovaries caused menstruation were already apparent in William Cullen's work in 1783 (see below). The advocates of uterine theories of menstruation had to contest this ovarian primacy to be able to explain menstruation as a process of any importance in the occurrence of pregnancy. As Johnstone wrote in 1887:

> The uterus, instead of being a mere appendage to the ovary, is as much a specific organ as the ovary itself.... Its association with ovarian activity is that of two separate departments of an army, each of whose work must be thoroughly accomplished before the one common object can be attained. They are both controlled by branches from the sympathetic system, and instead of their actions being

determined by each other, their orders come from that higher power which controls all functional activity (390).

The discourse which equipped Johnstone to discover complex functionality in menstruation was neurology, specifically the notion of the regulatory influence exercised by a distinct nervous mechanism seen to control general homeostasis, which Marshall Hall had first described as a distinct system in 1837. As the term 'hormone' would not be coined until 1905 to describe internal secretions with certain specific effects, in the nineteenth century the influence of the progesterone produced, following ovulation, by the corpus luteum (which it produces to sustain the uterine epithelium and then stops producing to cause it to be shed) was ascribed instead to nerve force.

From early to modern neurology

Before the rise of modern neurology, early eighteenth-century mechanistic philosophy had produced acutely mathematical explanations of menstruation, most notably in the shape of John Freind's *Emmenologia* (1703). Freind explained that the basic principles of physics entirely explained the 'menses'. In line with Galenic medical thought, Freind understood that the female body routinely produced an excess of blood with which to nourish the foetus in the uterus (or to be turned into milk when breast-feeding), and when not pregnant a mere thirty days of building up this excess nutrition would be enough to raise blood pressure and velocity sufficiently to cause the capillaries supplying the uterus to rupture (because these capillaries were, in Freind's view, the thinnest and most crooked capillaries in the female body, had the least insulating fat, and contained valves the least capable of locally equalizing pressure differences). The "infinite number of Accidents happen[ing] daily which interrupt this falutary work of Nature" were to be combated with the use of emmenagogues to bring on the menses, "fince almoft all the Diftempers, with which the Women are afflicted, are derived, as Hippocrates well obferves, from some irregularity of the Menfese" (ii, vi). Insisting that there was no poison or toxin in menstrual blood, Freind explained that disorders arose from suppressed menstruation because this general fullness reached dangerous levels. The immediate consequences of a suppressed menstruation included uneven blood consistency, a slower and weaker pulse and an unequal heartbeat, which would lead to inflammation of arteries, tumours caused by humours being retained in cracks created by the heavier blood, breathing becoming more difficult because of stagnation of the blood and heart palpitations because of increased pressure (85–102). In this model of menstruation, the nerves exercised no control. They were merely one of the organs impaired by this increase in blood pressure, in such instances as swollen veins constricting adjacent nerves and causing "a ſmaller influx of Spirits [through the nerves] into the Fibres of the Heart", in turn causing increased viscidity of the blood pumped out of the heart (82), or an uneven distribution of nervous liquid disrupting the balance of locations where two muscles are antagonistically arranged by causing one to contract constantly (his explanation of hysteria [106]). Notably, in Freind's model, the factors contributing to menstruation were those of deficiency rather than difference, including women's purportedly deficient perspiration, deficient vessels and deficient valves. He mentioned, for example, that menses could occur even if this monthly plethora did not come about because women in ill health may have "a too great weaknes of the Veſſels", or because of "the acrimony of the Humours" (42). In addition, as he defined it, the plethora of blood causing normal menstruation was not yet a phenomenon occurring exclusively in the female body. The male body also produced a plethora through excess concentrations of nutritive juice, its larger vessels possessing higher tolerances for increased blood pressure and velocity, and men's employment in physical labour meaning that they secreted much of this excess juice as sweat.

In *The English Malady* (1733), George Cheyne concurred with this understanding of nervous operation as a simple matter of physics, explaining that all nervous distempers arose from "the Want of a ſufficient Force and Elaſticity in the Solids in general, and the Nerves in particular, in Proportion to the Reſiſtance of the Fluids, in order to carry on the Circulation, remove Obſtructions, carry off the Recrements, and make the Secretions" (14–15). An increase in viscidity or salinity of nervous fluid, or a decrease in the elasticity of the nerve tubes channelling them (both of which would result from a rich diet), caused the nerves to stop functioning properly. A nerve had to be anatomically affected for a distemper to be 'nervous'. In this extensive account of the various flaws that would cause nerves to function poorly, so in turn causing "whatſoever from Yawning and Stretching, up to a mortal Fit of an Apoplexy" (14–15), Cheyne made only *one* mention of a sex-specific aetiology (when discussing fits or paroxysms): "If … Convulſions happen to the Younger Part of the Sex about a certain Time of their Lives (as they often do) then they generally proceed from ſome Diſorder in that great Affair, which ought, if poſſible, to be set to Rights" (222). In Cheyne's time, the uterus was still not seen as sufficiently linked to the nervous system to be capable of directly influencing the sensory nerves distributed to it. It influenced nerves only by throwing some substance into the body that increased the viscosity of nervous fluid or lessened the resistance of nervous tubes.

In contrast to Freind and Cheyne, for whom a nervous disorder arose from anatomically affected nerves, for Robert Whyte, in his November 1764 preface to his major treatise on nerves, disorders should only be called nervous "in ſo far as they are, in a great meaſure, owing to an uncommon delicacy or unnatural ſensibility of the nerves, and are therefore obſerved chiefly to affect perſons of ſuch a conſtitution" (iv). In contrast with Cheyne's model, where the occurrences disordering nerves would also disorder blood vessels, for Whyte there was a class of disorders which deserved the name 'nervous' because they *only* influenced the nerves: "We do not call the toothache a nervous diſeaſe, because the nerves of the teeth are greatly pained; but if, from a particular delicacy of conſtitution, the patient is, by this pain, thrown into convulsions and faintings, we call theſe ſymptoms nervous" (94).[2] Whyte listed the sympathies in which the uterus participated:

> [T]he vomiting which generally accompanies an inflammation of that organ, the nauſea, and depraved appetite after conception, the violent contraction of the diaphragm and abdominal muſcles in delivery, the headach, and the heat and pain in the back and bowels about the time of menſtruation, are ſufficient proofs of the conſent between the uterus and ſeveral other parts of the body (28).

The earlier menstrual model still influenced his thinking, as menstruation was one of his two exceptions to his principle that disorders in one part of the body are 'communicated' to others *only* via nerves (the other was the headache caused by wearing painful shoes). "May not the complaints of the ſtomach and bowels, from a ſuppreſſion of the menſes, and ſoon after conception," he wrote, "be owing not only to a particular sympathy between their nerves, but partly alſo to the change made in the quantity of the blood thrown upon theſe parts, by the obstruction of the uterine veſſels?" (77)

In rubbishing an earlier understanding of nerves as working poorly because their power to pump the nervous fluid was, through some accident, impaired, and replacing it with an understanding of nerves as working poorly through either a predisposition to weakness or though over-work, Whyte laid the basis for specialist and popular models of nerves that saw the female body as possessing a universally weaker nervous system. Nonetheless, he also helped to lay the basis for future neurologists to class all operations of the female body as the non-pathological actions of a purposeful system. Thus, while he observed that "[w]omen, in whom the nervous system is generally more move-able than in men, are more ſubject to nervous complaints, and have them in higher degree", he also insisted, for example, that "it is to be obſerved, that in women, the ſymptoms commonly called hyſteric, are leſs frequently owing to the unſound ſtate of the womb, than to faults ſomewhere elſe in the body" (118, 106–7). Although, for him, excessive menstruation might impair nervous power through diminished blood quantity, and while there might still be some morbid humour expelled by menstruation, whose retention would also impair nervous power, nervous control enabled him to dispute the existing plethora theory: "it is probably that the menſtrual evacuation is not owing to a general plethora, or increaſe of the maſs of the blood at the end of every month, but *to the particular ſtructure of the womb*" (177–8, emphasis added). This was one of the earliest statements that menstruation was a function of the uterus.

As neurology began to promote nerves to the position of the orches-trating network in charge of all operations of the human body (see Introduction), the workings of "the particular structure of the womb" would be aligned with nervous control. George Prochaska wrote in 1784 that "the structure and functions of the nervous system should be well understood by those who would determine what should be ascribed in animal actions to the operations and structure of the nervous system, and what should be clearly assigned to the immaterial soul alone" (364), so entertaining the possibility that the nerves might underpin *all* bodily events. Prochaska summarized what he saw as the first neuro-logical explanation of menstruation, Wolter Forsten Verschuir's thesis at the University of Groningen in 1766. As anatomists had noticed no evidence of an increase in the size of the uterus that either a uterine plethora or a general plethora would create just prior to menstruation, it was manifest

> that the menstrual blood is not contained in the uterine vessels pre-viously to menstruation, but is derived to the vessels and cavity of the uterus at the time of menstruation, and this by means of the nerves, which seem to be irritated by some stimulus as yet unknown, recurring periodically, and thus produce the derivation of blood to the uterus. Probably it is some latent peculiarity of the *vis nervosa* [the 'agency' possessed by the nerves], which recurs periodically (415–16).

In the neurological schema, menstruation appeared as a controlled process rather than a hydraulic excess. As early as 1777, when John Augustus Unzer claimed that "secretion and excretion may be simply direct nerve-actions of external impressions" (249), a schema was emerging in which menstruation could be reinvented as linked to a stream of internal and external stimuli, and for some the blatant inaccuracy of the plethora theory of menstruation supported this. Even William Cullen, who in 1783 still believed in the routine incidence of (albeit local or 'topical') plethora, and argued that "the flowing of the menſes depends upon the force of the uterine arteries impelling the blood into their extremities, and opening theſe ſo as to pour out red blood" (38-9), insisted that a plethora could not explain why women did not menstruate until about fourteen years of age. Instead, he ventured,

> [a]s a certain ſtate of the ovaria in females prepares and dispoſeſ them to the exerciſe of venery, about the very period at which the menſeſ firſt appear, it is to be preſumed that the ſtate of the ovaria, and that of the uterine veſſels, are in ſome meaſure connected together; and, as generally ſymptoms of a change in the ſtate of the former appear before thoſe of the latter, it may be inferred, that the ſtate of the ovaria has a great ſhare in exciting the action of the uterine veſſels, and producing the menſtrual flux. (43)

Although Cullen did not offer any explanation of the nature of this link between ovaries and menstruation (which he was one of the first to postulate), his most readily available explanation was neurological.

For example, it was noted that if menstruation was caused by a plethora of certain parts only, such accompanying 'symptoms' as swollen breasts could not be caused by this plethora. Rather, as Thomas Trotter asserted in 1807, "[p]ain and fullness of the mammae in some women usually accompany the period.... Such affections often arise from pure nervous sympathy" (195). In 1809 Bartholomew Parr wrote that a "topical load, in a system so irritable and so generally sympathizing as that of the uterus, would produce equal uneasiness; from the peculiar sympathy between the uterus and the breasts, the mammæ would swell" (189). In the face of new ideas of general nervous communication (see Introduction), the various 'symptoms' of menstruation were coming to be seen as caused not by any plethora of blood in either the uterus or the body in general but by the sympathy of other organs with the irritation of the nerves of the uterus, an irritation caused by the menstrual discharge occurring there, whatever its cause. In addition to this neurological account of accompanying 'symptoms', Cullen's explanation was also amenable to a neurological account of how the uterine blood vessels came to expel the excess blood in the first place. As Parr asserted, when menstruation became 'suppressed', this was not because of a deficiency in the amount of blood that the body produces, but because shocks to the nerves, such as cold, fright, falls and anxiety, could temporarily diminish the irritability of the uterine blood vessels. Diminished irritability also explained menopause for Parr. Although the blood might be supplied by a monthly plethora, it was only expelled by a nervous cause: the irritation of uterine blood vessels (191–2).

At the turn of the century, the idea that the uterus was especially irritable also began to give way to the principle of the irritability of the nerves distributed to it. In his 1800 Croonian Lecture to the Royal Society, Everard Home showed that surgically severing a nerve would lead to symptoms as severe as any that might originate at the end of the intact nerve (20–1). His successors confirmed this principle of universal irritability. In 1837 Benjamin Brodie wrote that the same was the case for efferent nerves: "an impression made either at its origin, or any where in the course of the trunk of a nerve, will produce effects which are rendered manifest, where the nerve terminates" (2–3). As nerves came to be seen as sensitive along their whole length, there was now much more reason to suppose a nervous disorder to be a consequence of irritation of the nerves alone rather than irritation of tissue merely communicated by that nerve. In the first decade of the nineteenth century, for example, the term 'neuralgia' was coined to denote pain restricted solely to the nerves. Pain at the extremity of a sensory nerve or spasms at the extremity of a motor nerve were to be seen as consequences of the nerve malfunctioning, and this malfunction in turn originated, Brodie argued, not just in damage or pressure to a nerve fibre but in such non-anatomical derangements as disorder of the digestive organs, and even "some derangement of the general health" (20, 27).

Given that menstrual pain could be explained as neuralgia produced by either irritation of the uterine nerves or the malnutrition of nerves caused by menstrual blood loss, the nervous system could be used to underwrite continuing ideas of female debility. Trotter, in his 1807 work, expressed the contemporary promotion of nerves to a first cause, writing that, in the bodies of sufferers from nervous diseases, while "the stomach, liver, pylorus, intestines, mesentery, kidnies and bladder, and the uterine organs in the female, have at times been found diseased; ... such deviations from the healthy state, must in general be secondary symptoms, the effect, not the cause, of these ailments" (213). The uterus would be disordered because the 'female constitution' was "furnished by nature with peculiar delicacy and feeling... and easily acted upon by stimuli" (52–3). It was this delicacy that caused nervous diseases to "fall mostly on the fair sex" (271). For example, he noted that "[n]ervous females particularly suffer at the period", even arguing that, because of this, gossip during menarche could severely damage health (228, 275). In spite of these assertions of female debility, however, Trotter did not advise any medication for those nervous disorders that he saw as caused by menstruation, partially because to tie menstruation to nervous control was also to insist on its normalcy. Even Brodie insisted that the hysteria he saw as the real pathology behind complaints of joint pain was just as possible for men as for women, because "[i]t belongs not to the uterus, but to the nervous system" (46), a result of mental anxiety as much as injuries, illnesses or loss of blood.

That hydraulic conceptions of the body's functions were fast receding in the face of neurological conceptions is evidenced by the frequency with which medical writers were now warning against employing the lancet in cases of suppressed menstruation. In Robert Morris and James Kendrick's *Edinburgh Medical Dictionary* of 1807, although they endorsed William Cullen's account of menstruation as a haemorrhage produced by topical plethora, they warned that bleeding would be ineffective. Instead, "restore the body to a healthy state, and this, as a natural excretion, will succeed" ('Chlorosis', n.p.). In 1816 John Reid wrote that "[t]o draw blood from a nervous patient is, in many instances, like loosening the chords of a musical instrument, whose tones are already defective from want of sufficient tension" (215). Advice on bringing on suppressed menstruation began to centre on methods for stimulating the nerves of the uterus, and the 1820s saw the emergence of the technique of injecting ammonia directly into the wall of the vagina to, as one practitioner wrote in 1824, "produce a sensation of orgasm" (Lavagna 166). This intervention was used in place of "[t]he remedy which is most efficacious" (i.e. masturbation), which "cannot, in many cases, be recommended without offending modesty, morality, and religion" (Lavagna 268). Menstruation, it was now being asserted, could not be established by internal medicines. Instead, it was necessary to stimulate the proper operation of the nerves.

By 1825 even those who clung to a belief in the occurrence of plethora were coming to see plethora as itself the product of nervous sympathy (e.g. Armstrong 289). Another theory of menstruation, from 1827, again rubbishing the notion of plethora, described menstruation instead as serving to 'soothe' the uterine irritation that was seen to generate female desire, again ascribing its cause to nervous signals (Abernethy 789). In 1829 James Blundell explained that it was "during the child-bearing period of life only, that the [menstrual] discharge flows, being, therefore, most probably associated, in the way of cause and effect, with aptitude for impregnation" (481). This now quite commonly postulated link of some kind between menstruation and the newly discovered process of ovulation expressed the perceived normalcy of menstruation now that it was seen to be the subject of nervous control. So while, in the 1840s, histological studies were expressing the rise of physiological attention to function by considering an organ's functions to be a property of its various component tissue types, neurology was offering a rival way of seeing menstruation as a property of the constant communication now coming to be seen to characterize the nervous system, particularly in the case of the reflex system.

The reflex and the uterus

The discovery in the 1830s that internal balance and sphincteric tone were controlled by a separate nervous system, distinct in its unconscious dependence on stimuli from both the (higher) volitional and (lower) vegetative nervous systems, both further impelled neurological accounts of menstruation and further lessened ideas of female constitutional debility. Marshall Hall's account of this new system (see Introduction) made it possible to conceive of the nerves as an avenue for non-pathological organic effects from organic causes. His 1837 treatise placed the uterus under the control of this 'excito-motory' system (as it was then called), explaining that "[t]hese excitor nerves may be viewed as guards of the orifices and exits of the animal frame", meaning that they included the afferent and efferent nerves of the eye, nostril, ear, pharynx, larynx, bronchia, cardia, ureter, gall-duct, rectum, bladder, vesiculae seminales and uterus. It was a reflex, he reckoned, that caused the fallopian tubes to grasp the ovaries just before ovulation (79, 93). Mid-nineteenth-century neurology explained menstruation with regular recourse to this non-sexed nervous system, rather than with recourse to femaleness. In the early 1860s John Chapman investigated the capacity of hot and cold shocks to bring on or halt menstruation, which showed, for him, that "the circulation of the blood in the womb is subject to the controlling influence of the sympathetic nervous system". He noted that, contrary to expectations that the uterus was susceptible to pathological functioning, this controlling influence meant that "the so-called functional diseases of that viscus are in reality abnormal conditions of the nervous ganglia which control it" (v). Between 1868 and 1884, William Ord carried out a number of studies that proved, for him, the controlling influence exercised by the sympathetic nervous system on the uterus. He explained that cases where menstruation was accompanied by inflamed and painful joints had long been unsuccessfully treated on the basis of the assumption that both pathologies were the result of a general rheumatic inflammation. Taking, instead, the view that reflexes in the spinal cord could 'communicate' pain and irritation from one area of the body to distant areas of the body, Ord treated both pathologies by 'subduing' the painful menstruation, which, he claimed, also caused the inflamed joints (as well as such symptoms as acne, sallow complexion and general ill-health) to disappear (20–1). This proved, for him, that, "by the morbid agency of the nervous system", the sensations attending the operations of the uterus were distributed to other parts of the body: "these organs [uterus and cervix] have the power, through centripetal nervous influence, of producing enormous excitement in the spinal cord" (4, 29). In both men and women, Ord argued, rheumatoid arthritis was the work of reflex nerve influence, and women suffered from arthritis more than men simply because they could experience one of its precipitating causes (excessive or painful menstruation) that men could not (9–11). The ascendency of the nerves deposed earlier explanations of the occurrence of arthritis in some menstruating women, such as Robert Todd's theory, put forward in the 1840s, that they were the result of "the unhealthy [internal] secretions of the uterus", which some had suggested was a virus and others lactic acid (qtd. in Ord 27, 28–9).

Moreover, while Hall and his corroborators argued that reflex arcs in the spinal column never troubled the supposedly distinct cerebral matter (and could not therefore be influenced by consciously experienced impulses or sensations), they came to be faced with several concerted arguments. First, as Johann Gaspar Spurzheim had been asserting in the 1820s, and as a number of his followers continued to stress as they developed the idea of a scale of developed species, the brain was a supplement of the nervous system – an upward growth – rather than vice versa, meaning that its constituent matter was not qualitatively different from the rest of the nervous system. Second, even if the three nervous systems (volitional, 'diastaltic' and vegetative) *did* function distinctly, they nonetheless communicated with each other in ways powerful enough to disrupt functioning. Spurzheim had written that "vegetative and phrenic [i.e. mental] life are mutually related; imperfect digestion, for example, disturbs the intellectual energies, and excessive

mental application, or moral sadness, interrupts the process of digestion" (195). Even Hall, in his June 1833 paper on the 'excito-motor'
system, had noted that

> One part of this inquiry is altogether untouched, – the influence of
> the mind and emotions, and the corresponding parts of the nervous
> system, upon the organs which are the subjects of the reflex func-
> tion.... Mental emotions modify the reflex function: they induce
> sickness, relax the sphincters: they also aggravate the diseases of
> this function, inducing the attacks of epilepsy, of the croup-like
> convulsion, &c (*Memoirs* 39–40).

Third, as David Ferrier pointed out in 1876, while reflex actions
explained why a patient with spinal injury might be volitionally paralyzed below the injury but still have her/his legs thrown into violent
convulsions if tickled on the feet, those without spinal injury could
also, as Ferrier pointed out, stop those convulsions caused by tickling
by concentrating hard (17–18). Michael Foster and Charles Sherrington
advised in 1897 that neurologists "must greatly hesitate to take it for
granted that the work which we can make the spinal cord or a part of
the spinal cord do, when isolated from the brain, is the work which is
actually done in the intact body when the brain and spinal cord form
an unbroken whole" (973).

As a consequence of these arguments, it began to be perceived that
conscious state rather than weak nerves could account for much menstrual irregularity. In addition, George Henry Lewes, at the end of the
1850s, typified new thought on the reflex when he refuted Hall's notion
that this nervous system, now known as the 'sympathetic', possesses "no
sensation, no volition, no consciousness, nothing psychical" (Hall 70).
For Lewes, the judgement it was able to show in responding to complex impulses meant that it constituted a consciousness of its own (2:
234–5; Ferrier 19–20). This meant that the reflex system could be influenced not just by the emotions of which it was, in Hall's model, the
seat, but by both the conscious thinking with which it was linked and
the unconscious thinking which it undertook. Consequently, menstruation's nervous symptoms could no longer be ascribed exclusively to
women's supposedly more intense emotional nature. The determining
structure of menstruation was the unconscious 'psyche' of the sympathetic system, and it was also under the influence of the conscious
psyche with which the sympathetic system was linked.

This neurological account of 'unconscious cerebration' – the term
proposed by William Benjamin Carpenter in his *Principles of Human*

Physiology (1844) – was readily applied to menstruation. One of the first of these was Carpenter's fellow proponent of the theory that the brain also contained reflexes, Thomas Laycock, who, in his 1840 treatise on the influence of the action of the ovaries on the female body, saw this action operating via nerves. Numerous studies, he argued, had shown that the ovaries and testes, rather than the uterus and penis, are the basis of sexual characteristics (removal of the latter had nowhere near the influence on sexual characteristics as removal of the former), but he also listed a number of examples from various species that showed that, even when neutered, animals and humans maintain such sex character-istics as superior size and strength in males (28, 77–88). This led Laycock to conclude that while the body's sex-specific characteristics do depend on the testes and ovaria, "the peculiarities of the sexes, whether mental or corporeal, including those referred to the generative organs, depend upon some particular organization of the primitive nervous system" (78). Menstruation was equivalent to haemorrhagic hysteria, Laycock argued, and hysteria unquestionably, for him, also occurred in men (79, 82). The controlling influence over menstruation was a nervous system that Laycock could not regard as decisively sexed. Citing evidence drawn from poetry, he insisted "that by universal consent the nervous system of the human female is allowed to be sooner affected by all stimuli, whether corporeal or mental, than that of the male" (76). He also strug-gled to describe this quantitative difference as a qualitative difference, explaining that "[t]he term affectability has been applied...to the con-stitution of woman, and it is both comprehensive and expressive" (76), but then going on to explain that this affectability arose from women's blood being slightly thinner (containing an average 78.7 per cent water to men's 76.7 per cent) (81).

Indeed, although Laycock agreed with a number of contemporaries who were proposing that menstruation was coincident with ovulation and so should be seen as equivalent to the onset of 'heat' in animals, he also disagreed with the common assumption that the ovaries cause menstruation: "we shall find that this periodic movement is not limited to the ovaria, but that it is an affection of the general system in which the ovaria partake; and that it is through these the secondary system in connexion with them is influenced, and all the attendant phenom-ena (those of menstruation) excited" (42–3). Nerve-reflex action was the foundation of nervous complaints attending menstruation:

it is universally acknowledged that irritation of the nerves of one organ may be communicated to those of a second having an anatom-ical or functional connexion. The catamenia are seldom established without aching and neuralgic pains of the back and lower extremities, partial anaesthesia (numbness,) and titanic contractions (cramps) of the legs. (116)

This occurred, he argued, because sensations passing through various cranial and spinal ganglia could cause sensations and increase and diminish secretions throughout the whole body (119). Thus could the actions of the uterus during menstruation produce general changes in bodily processes. But Laycock saw this periodic "affection of the general system" as normal in all species, claiming to have discovered periodic changes occurring in all animals every 3½, 7, 14, 21 and 28 days (60–1).

Admittedly, the normalcy implied by his treatment of these occurrences as 'phenomena' rather than symptoms did not extend throughout Laycock's work: "In healthy females the process is seldom inconvenient or painful", he wrote, "but in others it is accompanied by a variety of symptoms affecting the whole system. There are frequently hysterical symptoms of different degrees of intensity; the patient is whimsical, irascible, and capricious; is affected with vomiting, neuralgic pains in the head, face, side, or legs; and any disease to which she is subjected is at this period aggravated" (45). Nonetheless, a nervous account of menstruation permitted these nervous effects to be understood as symptoms only in women who were *already* unhealthy. Neurology provided the basis from which to derive a normalcy for menstruation which has so far been identified no earlier than the beginning of the twentieth century.[3] For example, Laycock's neurological model of menstruation anticipated Helen Warner's 1915 assertion that "[w]hen occurring normally, the symptoms [of menstruation] are few and light. There may be a sense of fullness, heaviness and slight irritability about the pelvic organs.... When menstruation is painful, when it is excessive or when it is diminished, it is an indication of disease, either local or general" (416).

Laycock also explained that, because reflex arcs existed in the brain, "the passions, the will, and other stimuli, acting upon the sensitive fibres of the brain," could produce effects analogous to reflexes (119): mental state could influence menstruation, in such cases as grief, fear, jealousy and rage suppressing the menses, and stimulation, for him, was the only way to restore them (174). Indeed, he saw nervous symptoms during menstruation as deriving partly from the irritation of the uterus caused by the heat generated by heavy clothing (142). As a contemporary wrote in the same year, uterine action did not of itself irritate the nerves. Rather, "the nerves of the uterus are unusually *susceptible of* irritation at the menstrual periods" (Bushnan, 99n, emphasis added).

In addition, according to the reflex theory the spinal cord *depended* for its operations on the influx of external stimuli. As Michael Foster and Charles Sherrington would later put it, "[t]he normal discharge of efferent impulses from the cord undoubtedly takes place under the influence of these incoming impulses; and it may be doubted whether the grey matter of the cord would be able, in the absence of all afferent impulses, to generate any sustained series of discharges out of its merely nutritive intrinsic changes" (993). As the constant influx of impulses was fundamental to the proper functioning of the organs under the control of the spinal cord, any disorder with these organs was liable to be caused not by deficiencies of that organ or the nervous system in general but by modifications in the impulses supplied by the external world.

Drawing on theories of unconscious cerebration, in 1887 James Oliver argued that the ability of the uterus to continue functioning if the spinal cord is severed did *not* mean that menstruation was beyond the influence of consciously felt impulses and sensations (379). Instead, the reflex actions that produced menstruation, although unfelt, were deeply immersed in the conscious mind. Explaining that "the physiological changes recurring from time to time in the uterus are anticipated by and in reality the sequence of a molecular disturbance arising spontaneously in some centre located in the higher parts of the cerebro-spinal tract", Oliver charted for menstruation a nervous cause that could be influenced by cerebration (379). "Like all other nerve centres fulfilling a similar dispensation, this uterine centre is undoubtedly beyond all volitional control," he agreed, "but is nevertheless capable of being disordered by emotional impressions" (379–80). He cited sudden shocks to the cerebral nervous system, which do no harm to the uterus itself, as the cause of immediate cessation of the flow. In turn, menstruation was also commonly the reflex cause of conscious aggravation (380). Oliver even explained that he disagreed with the theory that menstruation was a shedding of the uterine epithelium because he could not support a theory that saw any tissue type as so predisposed to such rapid degeneration. The neurological body, for him, operated "without degenerative change" (384). The uterine theorists of menstruation inherited this neurological account. In 1889 Robert Lawson Tait, having discredited, in his view, the theory that ovulation causes menstruation, wrote that, as to the action *really* causing menstruation, "I cannot assert anything in answer to this positively, save that it must be by some nervous mechanism, ... as is the contraction of all involuntary muscular organs." Indeed, a nerve mechanism controlling menstruation explained, for Tait, why "in some cases ... removal of both ovaries, both [fallopian] tubes and five-sixths of the uterus will fail to arrest menstruation" (320). He even claimed to have found this particular nerve trunk, and that destroying it caused menstruation to cease completely.

Menstruation and modernity

At almost precisely the same time that the modern meaning of 'nervous' entered English (see Introduction), so did 'menstruation'. In his 1765 work, Thomas Whyte was one of the first to use the new word 'menstruation' as an alternative to the older terms 'menses' (classical Latin for 'monthlies') and 'catamenia' (classical Greek for 'according to the month') (28). The adoption of this new term, meaning 'the production of the 'menstruum' or monthly discharge', was significant, as it referred to a process rather than a substance. Cathy McClive has shown that in French, *menstruation* (with the same meaning as in English) first began to be used by doctors as early as the 1740s, first appearing in print in Jean Astruc's *Traité des maladies des femmes [Treatise on Women's Diseases]* of 1761 (77). In English, Robert Hooper still used 'Catamenia' in his *Compendious Medical Dictionary* of 1798, but, by 1807, in Morris and Kendrick's *Edinburgh Medical Dictionary*, the term was 'menstruation'. By 1824 'menstruation' was a standard term. This alteration was highly significant. Although haemorrhagic, menstruation was now a deliberate process, only the outward signs of which were the emission of blood. For Morris and Kendrick in 1807, in an account that closely followed the medicine of Cullen, menstruation was associated with, but *not* synonymous with, other symptoms: "The eruption is generally preceded by symptoms that indicate its approach; such as sense of heat, weight, and dull pain in the loins; distention and hardness of the breasts, head-ach, loss of appetite, lassitude, paleness of the countenance, and sometimes a slight degree of fever" ('Menorrhagia', n.p.). By the time of Laycock's *Treatise on the Nervous Diseases of Women* (1840), the emission of blood was now one of a number of linked phenomena collectively known as 'menstruation':

> The phenomena of menstruation...consist in tumour of the mammæ, with a darker tint of the areolæ, weight and irritation about the pubes, pain in the loins, yawning, fastidious appetite, nausea, and not infrequently a sense of tension in the muscles of the neck, headache, and alternate pallor and redness of the cheeks: in addition, there is a flow of a sanguineous fluid from the vagina, varying in quantity from one to eight ounces. (45)

Thus, although 'menstruation' was supposedly changing in the 1880s to mean 'the denudation of the uterine epithelium, only *accompanied* by a discharge of blood', the term already referred to a process to which the discharge of blood was a mere addendum in the 1840s. Even James Oliver, who in 1887 was disagreeing with his contemporaries' proposal that menstruation was a shedding of the uterine epithelium, considered menstruation as much more than a simple venting of blood: "As the initial establishment and each subsequent recurrence of this monthly phenomenon is frequently accompanied by symptoms of a general as well as local character, we shall designate under the appellation *menstruation* the whole essential train of events, and not its mere outward manifestation" (378). The increasing sense of normalcy implied by this term did not follow from any secure knowledge of its function, since even in the 1880s there was no widespread consensus about the purpose of menstruation. Rather, confidence that every visceral event was a product of neurological control encouraged the idea that monthly vaginal bleeding, although, in Tait's words, "almost of itself a morbid process" (305), was just the visible aspect of a more extensive internal process that promised imminently to make much more sense.

Even before the 'discovery' of the purpose of menstruation in the mid-1880s, it was viewed as routine enough a function to be studied as one of the general monthly cyclical variations occurring in the human body, variations which some, including Laycock, were pointing out occurred in both male and female bodies. In the early 1870s, for example, Mary Putnam Jacobi found that the amount of urea excreted increases a few days before menstruation, decreases during menstruation and is at its lowest level just afterwards; that body temperature rises just before menstruation and falls during flow (while still staying above the inter-menstrual level); and that arterial tension rises from a minimum point just after menstruation to a maximum point just before, and is rapidly reduced during flow. William Stephenson, in his 1882 answer to Jacobi, substantiated the existence of these waves, but argued that they were not 'symptoms' of menstruation. Rather, menstruation was "associated with a well-marked wave of vital energy", merely another manifestation of this wave, along with variations in body temperature, daily weight of urea excreted and pulse rate, as it did not occur at the apex of *any* of these other wave forms (9–10). He also added that he had found the same wave forms in men (10).

This sense of normalcy derived not just from neurology but from the larger medical discourse of the era – physiology, with its commitment to studying phenomena in terms of function rather than form, and its attendant tendency to see all observable phenomena as potentially the result of systemic processes that were not, as yet, apparent. In 1885 Michael Foster cited Claude Bernard's mid-century discovery that the liver manufactured and stored glycogen, rather than simply secreting bile, as the major legitimization of physiology's search for tissue functions that were *not* apparent from anatomical structure (11). One of the ways that physiology, in its infancy, influenced theories of menstruation was in drawing attention to the possibility that the 'menstruum' might not be blood. Menstrual blood, after all, does not coagulate on contact with air, and although we know now that this is because of an anti-coagulating enzyme contained in the endometrium, in the early nineteenth century this was taken as evidence that the 'menstruum' was a deliberate secretion rather than (in the view of late seventeenth- and early-eighteenth-century mechanistic medicine) any simple voiding or habitual leaking of blood. James Blundell wrote in 1829 that "[t]he discharge, though of red colour, does not consist of blood; for though small concretions are now and then observed, yet, in the main, it is not found to coagulate, so as to form clots, or so as to harden the textures which are imbued with it" (481). In the 1820s and 1830s early champions of the idea of a dominant influence exercised by the ovaries over the female body described the substance as a deliberate secretion of the ovaries (Cabanis 280; Vaughan 206–12; Lee 'Ovaria'). In addition, although sixteenth- and seventeenth-century accounts of the 'menstruum' as an excretion of excessive nutrients or toxic matter had implicitly explained why the substance did not clot, from the viewpoint of physiology the substance was more likely to be an active secretion than a convenient venting point for body-wide excesses or toxins.[4] In 1827 Charles Scudamore wrote that, in medicine in general, "[t]he term secretion is preferable to the ancient term, humor, since it significantly refers to function" (5). While ideals of women's higher constitutional neurological disarray had been underwritten with reference to the starving influence of menstrual blood-loss on nerve fibres, if the substance was not blood then it did not represent such a loss. In addition, it also came to be seen that even if the 'menstruum' *was* blood, no nervous control would cause the body to constitutionally do itself harm. Morris and Kendrick's 1807 textbook, for example, was one of the first to note the emergence of claims that vaginal bleeding during infancy, pregnancy, lactation and old age was not menstruation, strictly defining menstruation as "the periodical discharge of blood from the uterus of healthy women who are not pregnant, or who do not give suck" ('Menstruation', n.p.). As a physiological process, menstruation had to be a definite secretion linked to certain, admittedly as yet unknown, internal bodily events.

Although eighteenth-century neurological claims that the body's automatic functions were controlled by sentience (e.g. Whyte 72–3) were soon surpassed by accounts that located the controlling 'will' of the viscera in discrete nervous systems, those neurologists arguing, against Marshall Hall, that reflex arcs could be influenced by conscious impulses or sensations produced a far more enduring explanation of a method for mental state to influence 'deep' functions (albeit unconsciously). Even though a woman's nerves were no longer held to be *anatomically* weaker than those of a man (the orthodoxy proposed during the early/mid-eighteenth century and typified by Friend and Cheyne), all nerves, in both men and women, were now seen to be *physiologically* purposed to distribute, throughout the body, the condition of the mind, which, in women, meant distributing the various supposed weaknesses of the female mind. In 1915 Helen Warner explained that "[s]ome women have nervous systems extremely sensitive to impressions, especially from the sexual organs. Such women are apt to be depressed and irritable during menstruation, and sometimes perhaps to have hysterical attacks at such times" (412). This was one of the ways that post-Renaissance medical discourses shifted from accounting for sex as a matter of quantitative dissimilarity to accounting for sex as a matter of fundamental and complementary qualitative difference. The neurological reinvention of menstruation was therefore no simple liberation of women from ideas of constitutional weakness. But the lack of any precise neurological account of sex did frequently produce models of menstruation that saw it as both normal and controlled by systems shared with men. For example, although Trotter saw women's nerves as constitutionally weaker than those of men, he also described a female body which was 'nervous' only in that it possessed certain organs, not possessed by the male body, which were liable to aggravate general nervous disorder. As 'nervous disorder' became 'the action of the nervous system', menstruation was neither pathological consequence nor cause of pathologies. Moreover, as menstruation came to be formally related to both conscious and unconscious cerebration, problems with, or associated with, the process, and thus women's general nervousness, were routinely traced to simple social anxiety, a product, itself, of nervousness.

While the plethora theory had seen menstruation as the result of a hydraulic phenomenon occurring in both women and men, and neurology rescinded this non-sexed uniformity in attaching menstruation exclusively to the ovaries and uterus, it nonetheless offered a new uniformity in establishing this link via the nervous system, whose functions it could not find to be intrinsically sexed. Although femaleness was quite widely pathologized in medical practice, neurology persistently discovered the female body to possess attributes also possessed by the male body, which medical discourse was not so eager to pathologize. It was therefore in a double sense that neurology contributed to the 'invention' of menstruation, establishing the infrastructure via which its cause (if not its purpose) was operated, *and* establishing a notion of nervous control that allowed menstruation to be seen as a routine bodily process in all women.

Notes

1. For example, Robert Lee's article on 'Ovaria' in the 1834 *Cyclopedia of Practical Medicine* claimed authorship of the theory that the flow of 'menstruum' is the human equivalent of the physiological changes undergone by animals in heat ('Finsburiensis' 129; Anon., 'On the Function' 294).
2. This alteration was key to the emergence of the new meaning of the term 'nervous' first noted by Johnson in 1755, and, in his 1764 preface, Whyte himself remarked that those conditions recently falling "under the names of Flatulent, Spasmodic, Hypochondriac, or Hysteric" had, "[o]f late, ... also got the name of Nervous" (iv). For more on this see the above Introduction.
3. For more on this see Strange (622–3).
4. For more on this see Stolberg (98).

Works cited

Abernethy. 'Physiological, Pathological and Surgical observations; Delivered in the Anatomical Course of Lectures at St Bartholomew's Hospital.' *Lancet* 186 (24 March 1827): 785–91.

Anon. 'On the Function of the Ovaries in Menstruation.' *Lancet* 1.899 (21 November 1840): 294–5.

Armstrong. 'On Chronic Affections of the Urinary and Uterine Organs.' *Lancet* 8.10 (10 September 1825): 289–93.

Blundell, James. 'Lectures on the Diseases of Women and Children, Delivered at Guy's Hospital. XXIX. Of Menstruation.' *Lancet* 2.307 (18 July 1829): 481–3.

Brodie, Benjamin C. *Lectures Illustrative of Certain Local Nervous Affections.* London: Longman, Rees, Orme, Brown, Green & Longman, 1837.

Bushnan, J. Stephenson. 'Observations on Dysmenorrhoea and its Treatment.' *Lancet* 1.893 (10 October 1840): 98–100.

Cabanis, Pierre. *Rapports du Physique et du Moral de l'Homme.* 4th ed. Paris, 1824.

Chapman, John. *Functional Diseases of Women. Cases Illustrative of a New Method of Treating Them Through the Agency of the Nervous System by Means of Cold and Heat.* London: Trübner & Co., 1863.

Cheyne, George. *The English Malady, or, A Treatise of Nervous Diseases of all Kinds.* London, 1733.

Cullen, William. *First Lines of the Practice of Physic.* Vol. 3. 1783. 4 Vols. Edinburgh, 1796.

Ferrier, David. *The Functions of the Brain.* London: Smith, Elder & Co., 1876.

'Finsburiensis'. 'Authorship of the Theory of Ovarian Menstruation.' *Lancet* 1.894 (17 October 1840): 129.

Foster, Michael. 'Physiology.' *Encyclopaedia Britannica.* 9th ed. 25 Vols. Edinburgh: Adam & Charles Black, 1875–89. Vol. 19 (1885): 8–23.

Foster, Michael, and C.S. Sherrington. Book 3. 7th ed. 1897. *A Textbook of Physiology.* 6th ed. 2 Vols. London: Macmillan, 1893–1900.

Freind, John. *Emmenologia.* 1703. Trans. Thomas Dale. London, 1752.

Everard Home. *On the Irritability of Nerves.* London: W. Bulmer, 1801.

Hall, Marshall. *Memoirs on the Nervous System.* London: Sherwood, Gilbert & Piper, 1837.

Jacobi, Mary Putnam. *The Question of Rest for Women During Menstruation, The Boylston Prize Essay of Harvard University for 1876.* New York: Putnam, 1877.

Johnstone, Arthur W. *Transactions of the British Gynaecological Society* (1887): 390.

Lavagna, Francesco. 'On the Efficacy of Ammoniacal Injections in Amenorrhoea.' *Lancet* 2.5 (1 February 1824): 165–9 & 2.8 (22 February 1824): 265–9.

Laycock, Thomas. *A Treatise on the Nervous Diseases of Women; Comprising an Inquiry into the Nature, Causes and Treatment of Spinal and Hysterical Disorders.* London: Longman, Orme, Brown, Green & Longman, 1840.

Lee, Robert. 'Ovaria.' *Cyclopedia of Practical Medicine.* London: 1837.

Lewes, George Henry. *The Physiology of Common Life.* 2 Vols. Edinburgh: Blackwood, 1859–60.

McClive, Cathy. 'Menstrual Knowledge in Early Modern France, c.1555–1761.' *Menstruation: A Cultural History.* Ed. Andrew Shail and Gillian Howie. Basingstoke: Palgrave, 2005. 76–89.

Morris, Robert, James Kendrick et al. 'Chlorosis', 'Menorrhagia' & 'Menstruation'. *The Edinburgh Medical and Physical Dictionary.* 2 Vols. Edinburgh, 1807.

Oliver, James. 'Menstruation – Its Nerve Origin – Not a Shedding of Mucous Membrane.' *Journal of Anatomy and Physiology* 21 (1887): 378–84.

Ord, William. *On Some Disorders of Nutrition Related with Affections of the Nervous System.* 1884. London: Harrison, 1885.

Parr, Bartholomew. *London Medical Dictionary.* London, 1809.

Prochaska, George. *A Dissertation on the Functions of the Nervous System.* 1784. Trans. Thomas Laycock. London: The Sydenham Society, 1851.

Reid, John. *Essays on Insanity, Hypochondriasis, and other Nervous Affections.* London: Longman, Hurst, Rees, Orme & Brown, 1816.

Scudamore, Charles. *A Treatise on the Nature and Cure of Rheumatism: With Observations on Rheumatic Neuralgia, and on Spasmodic Neuralgia, or Tic Douloureaux.* London: Longman, Rees, Orme, Brown & Green, 1827.

Spurzheim, Johann Gaspar. *The Anatomy of the Brain, with a General View of the Nervous System.* Trans. R. Willis. London: S. Highley, 1826.

Stephenson, William M. *On the Menstrual Wave.* N.Y.: William Wood & Co., 1882.

Stolberg, Michael. 'Menstruation and Sexual Difference in Early Modern Medicine.' *Menstruation: A Cultural History.* Ed. Andrew Shail and Gillian Howie. Basingstoke: Palgrave, 2005. 90–101.

Strange, Julie-Marie. 'Menstrual Fictions: Languages of Medicine and Menstruation, c. 1850–1930.' *Women's History Review* 9.3 (Spring 1996): 607–28.

Tait, Robert Lawson. *Diseases of Women and Abdominal Surgery*. Leicester: Richardson & Co., 1889.

Trotter, Thomas. *A View of the Nervous Temperament*. 1807. 3rd ed. Newcastle: Edward Walker, 1812.

Unzer, John Augustus. *Principles of a Physiology of the Proper Animal Nature of Animal Organisms*. 1777. Trans. Thomas Laycock. London: The Sydenham Society, 1851.

Vaughan, Walter. *An Essay on Headaches*. London, 1825.

Warner, Helen F. 'Diseases of Women and Children.' *Dr Chase's Combination Receipt Book*. 1915. Detriot: F.B. Dickerson, 1917.

Whyte, Robert. Preface. *Observations on the Nature, Causes and Cure of those Disorders which have been commonly called Nervous, Hypochondriac, or Hysteric*. 2nd ed. Edinburgh: J. Balfour, 1765.

3
Carlyle's Nervous Dyspepsia: Nervousness, Indigestion and the Experience of Modernity in Nineteenth-Century Britain

Hisao Ishizuka

Introduction

In 1895 Thomas Clifford Allbutt, the Regius Professor of Physic of Cambridge, inveighed against the widespread truism that the increase of 'nervous diseases' owed much to the decadent lifestyle of modernity. Allbutt argued that people said that nervous diseases were rife due to "living at high pressure, to the whirls of the railway, the pelting of telegrams, the strife of business" ('Nervous' 214), but he was not convinced by lay-people's naïve sociological diagnosis of the malady and denounced it as an error (215). That Allbutt, himself a strong supporter of neurasthenic diagnosis in Britain, had to deny the Beardian assertion that neurasthenia was the product of modern civilization (214) indicates how deeply the medico-cultural belief concerning nervous malady and modernity was entrenched in people's minds at that time. In fact, well before the arrival of the Beardian notion of neurasthenia in Britain, the British were believed to have experienced a train of nervous sufferings as the first citizens who had encountered the emergence of a commercial and industrial society. The British tended to see the Beardian malady, neurasthenia, as old wine in a new bottle (Sengoopta 98). From the eighteenth-century *English Malady* expounded by George Cheyne in 1733, who first linked nervousness and civilization, to Thomas Trotter's notion of the nervous temperament at the beginning of the nineteenth century, the link between nervous maladies and modern experience had been clearly established in British civilized life (Porter, 'Nervousness' *passim*).

Many historians, cultural and medical alike, have agreed that nineteenth-century Britain experienced an age of 'nervousness', in which people suffered from 'shattered nerves' – a euphemism for the diseases attributed to the functional disorders of the nervous system, such as nervous breakdown, collapse, prostration, irritability or depression (Oppenheim 5). Although it is true that Victorian people complained of 'nervousness', nervous patients of this era suffered not only from mental depression but also from somatic ailments of their digestive apparatus: stomach, liver and bowel. Regarding the experience of German and French neurasthenic patients, Radkau and Forth both suggest that disorders of digestion seemed to be a main concern of neurasthenic experience (Radkau 210; Forth 350). The idea that nervous disorders are fundamentally gastro-intestinal in nature seems to extend to nineteenth-century Britain, for the experience of gastro-intestinal complaints, variously called dyspepsia or biliousness, was in the forefront of the nervous experience of the modern era.[1] Victorians, in their daily lives, were more particularly preoccupied with chronic digestive complaints than with purely nervous (mental) collapse. The quotidian experience of digestive problems appears to be universal and even trivial; consequently, few cultural and medical historians have given it due attention.[2] In this chapter, I shall attempt to modify the common assumption that nineteenth-century Britain was dominated by nervous disorders, taking Thomas Carlyle's individual experience of dyspepsia as a case study within a history of nervous dyspepsia. In so doing, this chapter explores the modern experience of gastro-intestinal disorders in relation to the nervousness of modernity, focusing especially on 'nervous dyspepsia' as a disease of modern life.

Dyspepsia or biliousness as a fashionable 'Nervous' disease

One might state that the nineteenth century in Britain was a century of indigestion. But why particularly the nineteenth century? Disorders of the gut, such as indigestion, constipation and diarrhoea, have always been high on the list of patients' complaints. 'Dyspepsia', however, was discovered in the nineteenth century. Especially from about the middle of the century, medical books, learned and popular alike, were published on dyspepsia, indigestion, diseases of the stomach and the disordered stomach, steadily establishing themselves as a distinct genre (Gibbs 29, 38).[3] S.O. Habershon, one of the many stomach-doctors of the period, testified to the rapid upsurge of this genre as well as the rampancy of stomach diseases: "Although nearly every year new works have appeared on diseases of the stomach, still the maladies affecting this organ are so numerous...that there is ample scope for the records of individual experience" (13).

The rapid rise of this genre of dyspepsia in the nineteenth century owed much to the critical revision of indigestion as a primary disease, as distinct from an accidental symptom accompanying primary diseases such as pneumonia or flu. In the nineteenth century, dyspepsia advanced to being regarded as a 'disease' and a diagnostic entity, so that patients could identify themselves as 'dyspeptic'. Allbutt's lamenting the inflation of dyspepsia's diagnosis at 1884 as a disease captures well the mood of the time. In the first part of the Gulstonian Lectures on visceral neuroses, Allbutt set out to rectify what he perceived to be the vulgar error on dyspepsia which was widespread among both doctors and patients: " 'Martyrs of dyspepsia' are to be found at every street-corner, and are said to form something little less than the staple of those who drift from consultant to consultant" (*Visceral* 3). Medical journals and "drug-houses" (pharmacies and dispensaries), continues Allbutt, all co-operated in the production of "the demon of dyspepsia" (4); moreover, patients participated in the boom, declaring themselves as "old dyspeptics", and demanding "specifics" from doctors (5). Allbutt denounced them for mistaking symptom for disease: "dyspepsia...is not a disease, but a symptom" (5). Dyspepsia, he felt, did not deserve the name of a malady, so it should be reduced to the proper position of symptom common to many diseases (7). Allbutt's criticism reflects the extent to which dyspepsia circulated as a popular disease.

The fad of dyspepsia, intensifying from the mid-century onwards, had precedent in the disease of the liver known variously as biliousness, the bile or the bilious, which was another fashionable disease, especially in the first half of the century. Biliousness had already been fashionable among the higher classes flocking to Bath in the last decade of the eighteenth century ("nerves and nervous diseases were kicked out of doors, and bilious became the fashionable term" [Adair 6]),[4] but by the first decade of the nineteenth century it seemed to establish itself as a fashionable disease among the wider population. Biliousness was so-termed because a train of disorders such as indigestion, constipation, heartburn, low spirits, headache and insomnia, were deemed to be caused by a morbid condition of bile, in either quality or quantity; since the bile was secreted in the liver, the malady was also called "liver complaints", as in the title of the London surgeon John Faithhorn's 1815 work, in which he discussed the centrality of biliousness in chronic diseases. Seeing that "the prevalence of liver complaints" was so general,

Faithhorn was fully satisfied that the liver was "the chief seat of most" chronic diseases, and was convinced that "the grand source of health and disease" was "connected with…the liver" (33). Faithhorn's strong hypothesis that the liver was responsible for bodily welfare seemed to be shared by many popular medical writers of the time (Kitchner 242–3). Some medical writers reprimanded medical practitioners for the excess of biliousness diagnosis fashionable at the time. In 1819 T.M. Caton denounced those popular writers who "attached too much importance" to the liver and ascribed almost all diseases, including even "cancer, consumption, insanity", to "hepatic derangement" (11–12). Thomas John Graham also acknowledged the popularity of the term in 1828 ("the term 'liver complaints', is now in the mouth of every one" [2]), while he regretted that this "fashionable term" was too much abused (2–3). Formation of a popular medical hypothesis that all chronic (and sometimes some acute) diseases were attributed to the liver and the bile seemed to reach its height in the 1810s–20s.[5]

After the arrival of dyspepsia as a modish malady around the mid-century, biliousness, nevertheless, did not yield its place. In the latter half of the century, people still referred to their chronic malady as 'biliousness', and as Leared lamented in 1875, "[n]othing is more common than for people to say they are 'bilious'" (84). Worse, it was not only patients but also medical practitioners, who abused the term 'bilious'; they were daily attributing pains and various illnesses to the liver and found it "a convenient scapegoat" (245–6). Since both organs, the liver and the stomach, are mainly involved in digestive function (moreover, the structural position of both organs is very close), the manifestation of dysfunction of the digestive apparatus is the same either in dyspepsia or in biliousness. So, whether these disorders were to be called 'dyspepsia' (ill effects traced to the stomach) or 'biliousness' (in which the liver and the bile are culpable) seemed to depend much on the person's preference. Thus, the fad of biliousness and that of dyspepsia, a slight latecomer, formed, as it were, two sides of the same coin: the modish malady of digestive apparatus rampant in the nineteenth century.[6]

Dyspepsia or biliousness was not just a disease of digestion (i.e. indigestion), however; the symptoms of these diseases encompassed a series of psycho-somatic 'nervous' outcomes, including headache, nausea, palpitations of the heart, debility, gloom, irritability, despondency, giddiness and the dysfunction of the senses, similar to those of nervous or hypochondriacal diseases of the eighteenth century. In fact, nineteenth-century dyspepsia was a modernized version of the 'hypo' of the early modern era, a disease whose origin was originally attributed to the dysfunction of the belly. This fashionable malady, variously called, according to the authors, as 'hypochondriasis', 'melancholy' or 'spleen', was distinguished by its psycho-somatic 'nervous' symptoms similar to dyspepsia (e.g. Jackson 274–301). The continuity of 'nervousness' or 'hypo' and 'dyspepsia' or 'biliousness' is to be seen in the nervous manifestations of the disease and in the representation of the sympathetic influence between the brain and the digestive organs. It was widely believed that affections of the brain depended on disorders of digestive function; the brain as the *sensorium commune* was the principal organ to be affected by the disorders of the abdominal viscera, the greatest centre of 'sympathy', resulting in numerous kinds of mental dysfunction such as "confusion of thought, unsteadiness of the mind, irritability of the temper, defect of memory" (Johnson 83–4). This digestive derangement also affected sense perception, causing "deafness, vertigo", or "defect of vision" (84–6). Another medical writer on dyspepsia pointed out in 1846 that the dyspeptics were apt to be irritable and despondent, having little inclination to mental exertion due to the powerful influence of the stomach on the mind (Clark 15).

That the digestive organs had more powerful effects on the mind than one might expect also intimated that the stomach or the liver were not just the organs of digestion. John Faithhorn's early nineteenth-century view of the liver as the medium of the mind and the body is a case in point. Faithhorn explained that nervous disorders and hypochondriasis are produced by liver disorder through "the powerful influence of the liver, on the nervous system"(55); he further meditated on the cause of suicide, ascribing it to "a fault of the biliary secretion of a disordered liver", for if the liver, "that fine matter", serving as "the medium between the body, and the thinking part" is, on dissection, found to be diseased, "we must evidently refer it [suicide] to this source" (56). The digestive apparatus, for Faithhorn, formed a vital knot connecting the mind and the body.[7]

The sympathy between the brain and the digestive apparatus was not one-sided but mutually influential, as the brain, for its part, exerted a substantial influence on the digestive apparatus. While mental disturbances were affected by the morbid conditions of the stomach and the liver, indigestion was caused by the disturbance of 'nervous energy' of the brain. Any kind of excess of mental activity, such as intense thinking, prolonged mental anxiety and stress, which could easily exhaust the limited amounts of 'nervous energy' possessed by one person, caused the state of the digestive organs to deteriorate (Caton 20–2; 'Review' 336). For instance, any hard exercise, mental or physical, before breakfast would be harmful for healthy digestion, for such exercise would spend the nervous energy which should be reserved for the aid of the digestive process (Thomson 69, 187). It is important to note here that many medical writers asserted the primacy of the mental over the physical as the chief cause of indigestion. James Johnson was explicit on this point: "The operation of physical causes…dwindles into complete insignificance, when compared with that of anxiety or tribulation of mind" (95). Brinton also favoured "undue intellectual exertion" as the primary agency in gastric disorder, though he admitted that dyspepsia had no single and simple cause (260).[8]

Medical writers tended to subdivide dyspepsia into many categories – gastric dyspepsia, atonic dyspepsia, nervous dyspepsia and so on (Clark 13; Ross 211). Among these categories, nervous dyspepsia was said to be the most frequent complaint (Dewar 41). Nervous dyspepsia was so-called because indigestion was primarily caused by a weakness of the nervous system or the deficiency of nervous power, although the definition varied, depending on the writer. Having seen the ways that dyspepsia or biliousness were deeply related to 'nervousness', we may safely term the fashionable disease of the time as 'nervous dyspepsia'. Having said that, we should also notice that contemporary people did not call themselves 'nervous dyspeptics'; rather they preferred to call themselves simply 'bilious' or 'dyspeptic'. The term 'nervous dyspepsia' is used here as a heuristic word for illuminating Carlyle's dyspeptic experience.

Carlyle's nervous dyspepsia

Many Victorians were 'dyspeptic' or 'bilious'. Distinguished men of letters such as Thomas Carlyle, Charles Darwin and T.H. Huxley, were inveterate and self-confessed dyspeptics, and defined themselves as such. Huxley, for instance, referred to his life-long illness, commencing at about the age of 13 or 14, as a "constant friend, hypochondriacal dyspepsia" (Gould 109). Following the diagnosis of the day, he ascribed an inexplicable prostration to an "obstruction of the liver" (Gould 111). His strong association of mental depression with the liver confirmed the rise of nervous dyspepsia in the lineage of eighteenth-century hypochondria.

Huxley's obsession with the liver was also shared by Charlotte Brontë, another sufferer from dyspepsia ("headache and dyspepsia are my worst ailments" [Gaskel 310]). During the winter of 1852, Brontë felt extremely unwell, with a bad depression of spirits and an "internal congestion", with aching pain in her side and chest; sleeplessness, a loss of appetite and slow fever ensued (Gaskel 241–2). Having inherited a constitutional tuberculosis, Charlotte considered that the lungs were affected and feared it might be fatal. Consultation with a doctor, however, gave her immense relief, for he ascribed all her sufferings to "derangement of the liver"; consequently, her health suddenly returned: "My sleep, appetite, and strength seem all returning" (Gaskel 242). This example illustrates the placebo effect of a liver complaint (biliousness or dyspepsia was not considered a particularly detrimental disease) and shows the extent to which the belief in the liver was entrenched among Victorian people.

The popular writers of the day also responded to the rise of nervous dyspepsia, introducing a dyspeptic man as a stock character in their fictions. Wilkie Collins, for instance, in one of his popular mysteries, *The Dead Secret* (1857), featured Mr. Phippen as a "Martyr to Dyspepsia" (41, 61). Collins depicted Mr. Phippen as a dyspeptic sufferer *par excellence*. Being a good hypochondriac, he was always preoccupied with his stomach, what he ate, and what amount he was to eat, measured with a pair of apothecary's scales (41, 44, 46). Mr Phippen was willing to speak about the state of his digestion, his "clogged Apparatus" (46) or the condition of his tongue, a sign of dyspepsia, in mournful or "languidly sentimental tones" (41). The somatic condition intersected with the psychic one in Mr Phippen's body, for his loss of memory was ascribed by him to his indigestion (47). For Mr Phippen, the speculative view of the world depended upon the condition of the liver or the state of "biliary secretions", arguing that if one's "biliary secretions" were "right", one took "bright views", but if they were wrong, one took "dark views" (63) – a "bilious philosophy... tinged with bile" (63). Collins's comical representation of Mr Phippen as a sufferer of nervous dyspepsia was shared by other popular writers such as Charles Dickens, Sheridan Le Fanu and William M. Thackeray.[9] These examples show that by the mid-nineteenth century, the dyspeptic man entered the fictional world as a stock character.

Among Victorians, Carlyle was probably the most well-known literary dyspeptic. In his volumes of correspondence, Carlyle left invaluable materials on his dyspeptic life, meticulously recording daily concerns about and experiences of his digestive problems. Around 1818, Carlyle began to experience his lifelong dyspepsia and biliousness, with bouts of anxiety and depression.[10] Although in *Reminiscences* written in 1869 Carlyle confessed to being a long-suffering patient of dyspepsia ("my long curriculum of *dyspepsia*" [241]), more frequently he referred

to (and defined) his ill health as 'bilious', at least in the existing published letters. From a chronic weakness of the digestive organs of a
'nervous' kind in which 'bilious', 'dyspeptic' or 'nervous' were commin-
gled, through 'billus' – a term favoured by him and associated with the
diseased liver –, and finally to 'dyspepsia', Carlyle's shifting way of referring to the illness and identifying himself as such, closely followed the
medico-cultural history of naming of 'nervous dyspepsia'. Carlyle opted
for a self-fashioning of the illness according to the modish manner
of his era.

Whilst his principal pain was in the digestive apparatus, Carlyle very
often complained of mental and intellectual effects of illness, which
echoed the contemporary medical idea of the dreadful effects of the
diseased stomach or the liver on the mind. The disordered state of the
digestive apparatus brought about malfunction of the mental activities:
depression, inertia, low-spirit, gloom, and in Carlyle's favourite formulation, 'stupid(ity)'. One day, writing the second chapter of the *French
Revolution*, Carlyle was seized with "bilious humour", which manifested
itself "in the shape of *Stupidity*" (7: 96). At another time, on returning from a journey, Carlyle was rendered "utterly stupid...owing to
dyspeptic reasons" (29: 71). The record of his illness tells that Carlyle was more concerned with the degrading effects of dyspepsia on
the intellectual faculty than with the bodily pain itself: "It is not the
pain of those capricious organs; that were little; but the irresistible
depression, the gloomy overclouding of the soul, which they inevitably
engender, is truly frightful" (1: 362). Another 'nervous' aspect was
headache, popularly called "bilious headache" at the time (Brigham;
Faithhorn 54; Gould 147; Leared 33). Carlyle also very often complained of "bilious headache" (24: 186), especially about the meal he
ate which led him to have indigestion (2: 430, 20: 152, 21: 200) and
about insomnia, which he took as another sign of biliousness (14: 159,
24: 205).

Probably the most illustrative nervous symptom of Carlyle's 'billus'
was the disorder of his eyesight. During the difficult time when he
was penning *French Revolution*, Carlyle discerned in his left eye "a small
speck of *mote*" (8: 262), soon identified as "muscae volitanetes" (8: 287)
in medical terms. Interestingly, Carlyle associated it with 'billus'. Carlyle found the speck of mote after he was swamped by a "dark *green*
cloud of bile" that hung over him (8: 262). Later, he again associated
the bile with the "black speck" of the eye; just after he reminded John,
his beloved brother, that he was in the "valley of the shadow of deepest
Bile", he went on to remark that "the black speck...attends me pretty

constantly", and connected it with the liver: "Memento jecoris, jecoris [reminder of the liver]!" (8: 286). In a letter to Emerson, he reiterated the association, using a militant image: "a little black speck dances to and fro in the left eye (part of the *retina* protesting against the liver, and striking work)" (8: 336). Carlyle's association of 'muscae voluntantes' with 'billus' was at one with the contemporary medical idea of nervous aspects of dyspepsia or biliousness, according to which, dyspepsia (diseased state of the stomach or the liver) was said to pervert the sense of sight and hearing: "One not uncommon result of stomach disorder is the appearance of dark bodies like flies" before the eyes (Leared 53; see also Dewar 46; Holbrook 54; Johnson 86). This symptom was shared by dyspeptic "Mr. Phippen" of Collins's fiction, who also experienced "black bilious spots" dancing before his eyes (49).

Carlyle's hyper-hypochondriac lived experience of 'billus' or 'nervous dyspepsia' was psycho-somatic, with its special emphasis on 'nervous' symptoms of the chronic kind. It should be noted here that there was no substantial reference to the brain in Carlyle, whereas his reference to the 'nerves' or 'nervous system' was, if unevenly, persistent throughout his correspondence. Carlylean intellectual activity always drew its energy from the abdominal viscera – the liver, the stomach and the bowels. It depended on the state of the digestive faculty. After having roast beef (as his usual dietetic experiment), he triumphantly concluded that it was "not the character of being any great quickener of the intellectual powers" (2: 96, 106). At one point Carlyle even hinted at the possible increase of "nervous energy" due to a "great supply of bile" (17: 4) during his journey to Liverpool, which might suggest that he saw the liver (from which the bile was secreted) as a kind of reservoir of nervous power, not unlike the brain which was commonly assumed as such.[11] For Carlyle, nervousness came from the abdomen up to the noble parts, and not vice versa.

Experience of modernity: Nervous dyspepsia as a disease of modern life

In the early nineteenth century, 'nervous dyspepsia' seemed to pervade all ranks of people in England, as James Johnson announced:

> In civilized life, indeed, what with ennui and dissipation in the higher ranks – anxiety of mind, arising from business, in the middling class – and poverty, bad food, a bad air, bad drink, and bad occupations, among the lower classes, there is scarcely an individual in this

land of liberty and prosperity – in this kingdom of "ships, colonies, and commerce," who does not experience more or less of the "English malady" – that is to say, a preternaturally irritable state of the nervous system, connected with, or dependent on, Morbid Sensibility of the stomach and bowels. (102–3)

Echoing Cheyne's implicit celebration of "the English Malady" as a disease of a civil and commercial society, Johnson praised England for civilization, liberty and prosperity. Nervous indigestion, connecting to the "morbid sensibility of the stomach", was the penalty that civilized people paid (see also Dewar 9–10). As the century progressed, however, it became much clearer who specifically should pay for it: the middle classes. Johnson had already moved in this direction, since in the aetiology of nervous indigestion he was explicit in placing mental factors ("anxiety of mind") over physical ones, targeting "the middling class" as implied readers. As seen above, many medical writers of the time stressed mental causes over physical ones, and some writers put less importance on the organic lesion. Who was it, then, that most consumed and exhausted their mental energy and intellectual power, and who were often gnawed by mental toil and worry? Most likely, it was those who were engaged in some kind of "brain-work" (Brinton 261), which spent precious nervous energy by an "overworking of the nervous system" (Holbrook 66). And it was the middle class that was particularly, but not exclusively, associated with "brain-work", in sharp contrast to the labouring classes who performed physically demanding work, and to the upper classes who were deemed to lead an idle life. So, literary and professional men as well as businessmen ("the persons engaged in offices" [Leared 14]) were believed to be particularly prone to nervous dyspepsia (Aytoun 312; Leared 72; Thomson 296), which was specifically articulated as the disease of middle-class mental workers, or as *The Times* put it in 1884, citing the physician James Crichton-Browne's words, "the curse of brain workers" ('London' 15).

The emergence of nervous dyspepsia as a particularly middle-class disease connected with "brain-work" indicates that the implications of "civilized life" radically altered during the Victorian period. Before the nineteenth century, it was the privilege of upper-class people to enjoy a civil, polite and fashionable (i.e. civilized) life; supported by landed property, they led an affluent and idle life of ennui which made their nerves and fibres very lax, conferring on them 'the English Malady' – an honourable badge of their civility and modernity (see Cheyne *passim*). In the nineteenth century, however, modern civilized life was experienced particularly by middle-class (brain-)workers as one that would not allow indolence, a living in favour of the ideal of industry and the work ethic; civilized life strained nerves, rather than slackening them. It allowed no idle hours to middle-class men: *"Laborare est orare"* (Aytoun 321). In fact, the middle-class businessman had to work for a living, for a better living and for promotion. The rapidly advanced mode of modern industrialization and urbanization fostered "a spirit of emulation" and helped the "competitive system" of the society to grow to an extent "formerly unknown" (Leared 16; see also Aytoun 322). Accordingly, a large amount of the nervous energy necessary for a good digestion was used in "mental toil or business anxieties" (Leared 10), which led inevitably to dyspepsia. Unlike eighteenth-century nervousness, a disease of the *bon ton*'s idleness, nineteenth-century nervous dyspepsia was a disease of labour; it was the penalty that middle-class people had to pay for their blessed or cursed experience of the modern mode of civilized life.[12]

Carlyle was also aware of the ill-effects of the modern mode of life on his ever-shifting constitution. Carlyle's 'billus' rapidly increased after moving to London. He was relatively well into the Craigenputtoch years (1828–33), and he made it a custom to travel to Scotland for a change of air from the busy, stressful life of London. These facts indicate that Carlyle felt uneasy (bilious) with the nerve-wracking mode of life in London. The "insane hubbub" of modern life, along with the "velocity of all things", exacerbated his "poor excitable set of nerves", which could not withstand the "tear and wear of this huge roaring Niagara of things" (11: 24). The express train, a benefit and a symbol of modern technology, tortured Carlyle, giving him a severe headache which he called "railway headache" (25: 253), probably one symptom of "railway fatigue" widely experienced by railway passengers at that time: the constant mechanic vibration of the train was deemed to give serious effects on the nervous system (Harrison 248–9; Schivelbusch 134–70).[13] Carlyle's dyspepsia, then, was partly a product of the mode of modern life.

Like most middle-class businessmen in a competitive society, Carlyle was impelled to work and work more. And like businessmen who were made ill by overwork, Carlyle always became extremely "billus" after working too intensely (4: 66, 6: 251; 8: 213; 12: 252; 16: 227). At one time, he was wise enough to calm the hectic pace ("I did not overwork myself" [8: 202; 16: 61]). Abstaining from his work or taking rest, however, did not have a beneficial effect on him, because taking rest gave only a temporary relief, then further precipitated his bilious malady. For instance, after finishing the second volume of *The French Revolution*, Carlyle determined to have rest and spent idle hours for a week, but as soon as the second week began, "the *bile* declared itself," which had been "lying hidden" (8: 343). Being 'billus' or 'dyspeptic' after taking rest as well as after hard work – this was a pattern typical of Carlyle to be repeated throughout his life (e.g. 18: 122; 20: 19). Carlyle confessed one day that work of the past, present and future was "a weariness" to him (17: 53), but having an idle hour was all the more unbearable for him because it produced nothing ("Donothingism" [20: 65]); it compelled him to set to work, which he deemed the only "remedy" for his "billus" (17: 366). Sitting silently in his room and meditating on a new book was just "one way of recovering" from his biliousness (12: 252). Work, for Carlyle, was double-edged: it fostered his ill-health, while at the same time it was the only remedy for it. Fully internalizing the doctrine of the work ethic, Carlyle was always preoccupied and guilt-ridden with anxiety about idleness ("Donothingism ... press[es] heavy on my conscience" [14: 176]). Carlyle's nervous dyspepsia, ironically, was a product of his own work ethic, of which he was the most efficient spokesman of the time.

Under the pressure of the hectic mode of modern life and following the ideal of the work ethic, Carlyle's everyday life was almost always plagued with the sullen demon of 'billus', 'biliousness' and 'nervous dyspepsia'. The pervasiveness of his biliousness was like the black speck floating before his left eye, and it is no wonder that Carlyle later nicknamed the black speck the "French Revolution" (17: 198) which he deemed to be the "hardest work" (8: 5). For Carlyle, it was a cursed reminder of the demand of work within a perpetually nervous modernity.

Well before Beardian neurasthenia had diffused in Britain, British people, particularly middle-class workers who were able to avoid physical labour, had already experienced modern 'civilized' life with nervous dyspepsia. The detailed record of his dyspeptic or bilious experience in his correspondence illustrates the ways in which Carlyle's dyspeptic body was deeply implicated in the medico-cultural assumption of nervous dyspepsia. As an efficient promulgator of the work ethic, however, Carlyle's dyspeptic body not only embodied the end-result of his own work ethic, but also served to elevate dyspepsia as an honourable malady, a badge of hard-work (and, if not often, of success) in his fame as the most influential and industrious man of letters.

This chapter is dedicated to the late Toshikatsu Murayama, whose brilliant scholarly career was untimely cut short at the age of 38.

Notes

1. Part of the research of this subject was published in my 'Biographia Dyspepsia'.
2. See Bynum for a rare exception.
3. To list some examples of the titles chronologically, A.P.W. Philip, *A Treatise on Indigestion* (1821); Jonathan Hutchinson, *On the Forms of Dyspepsia* (1855); William Brinton, *Lectures on the Diseases of the Stomach and its Kindred Diseases* (1877); John Dewar, *Indigestion and Diet* (1878); William Roberts, *Lectures on Dietetics and Dyspepsia* (1885); and Adolphus E. Bridger, *The Demon of Dyspepsia* (1888).
4. Adair rebuked the assumption that the diseases called bilious were those of the upper classes; instead, he found these complaints were much common among the "soldiers, seamen, and day-labourers" (16).
5. Some examples are John Faithhorn's *Facts and Observations on Liver Complaints, and Bilious Disorders* (1815), Joseph Ayre's *Practical Observations . . . of Those Disorders . . . Strictly Denominated Bilious* (1818), J. Lynch's *A Practical Treatise on Nervous, Bilious . . . Affections* (1822), and the anonymous *A Physician's Advice of the Prevention and Cure of Bilious and Liver Disorders* (1824).
6. In the early twentieth century, Gould still recorded the popularity of the use of 'biliousness': "As to the lay world that word is used almost every day in almost every household in the land" (145).
7. Gibbs mentioned the non-digestive function of the stomach, which was commonly emphasized at the time (39n9). It is interesting to see that Gould recorded that neurasthenic cases were often explained by "derangement of the liver" via mental influences (151). It seems that the liver (or the bile) as something related with a nervous or sentient principle was deep-seated in the minds of Victorians.
8. See also Brigham (97) and an anonymous reviewer who, in 1849, remarked that "dyspepsia commences in the brain" ('Review' 327).
9. For example, Mr. Cobler in Dickens' 'The Boarding House' (1834), Mr. Jennings in Le Fanu's 'Green Tea' (1872), and Joseph Sedley in Thackeray's *Vanity Fair* (1847–8).
10. For a good biography of Carlyle, see Heffer.
11. As mentioned in note 7, the liver (or the bile) as having something like nervous energy seems to occupy some place in the Victorian mind. Carlyle's usage of "bile" is very interesting in that it reminds us of the activity of 'animal spirits' of the early modern era, which also suggests the link between 'hypo' and nervous dyspepsia.
12. This line of argument is partly indebted to Porter's concluding remark on the continuity of eighteenth-century nervousness and nineteenth-century neurasthenia ('Nervousness' 42), but also critically modifies it, for the vital factor of the continuity should be sought in "indigestion".

13. See also chapters by Jane F. Thrailkill and Shelley Trower in this collection for the impact of rail travel on the nerves and the ambivalent use of vibratory modern machines, respectively.

Works cited

Adair, James. *Essays on Fashionable Diseases*. London, 1790.

Allbutt, T Clifford. 'Nervous Disease and Modern Life.' *Contemporary Review* 67 (1895): 210–31.

———. *On Visceral Neuroses Being the Gulstonian Lectures*. London, 1884.

[Aytoun, W.E.] 'Meditations on Dyspepsia.' *Blackwood's Edinburgh Magazine* 90 (September 1861): 302–22; 406–19.

Brigham, A. 'Influence of Mental Cultivation in Producing Dyspepsia.' *Liver Complaint, Nervous Dyspepsia, and Headache*. By M.L. Holbrook. New York, 1876. 95–133.

Brinton, William. *Lectures on the Diseases of the Stomach*. 2nd ed. Philadelphia [London], 1865.

Bynum, W.F. ed. *Gastroenterology in Britain: Historical Essays*. London: Wellcome Institute, 1997.

Carlyle, Thomas. *The Collected Letters of Thomas and Jane Carlyle*. Ed. Charles Richard Sanders et al. Durham: Duke UP, 1970.

———. *Reminiscences*. Ed. K.J. Fielding and Ian Campbell. Oxford: Oxford UP, 1997.

Caton, T.M. *A Treatise on Indigestion*. London, 1819.

Cheyne, George. *The English Malady*. 1733. Ed. Roy Porter. London: Routledge, 1991.

Clark, James. *The Sanative Influence of Climate*. 4th ed. London, 1846.

Collins, Wilkie. *The Dead Secret*. Ed. Ira B. Nadel. Oxford: Oxford UP, 1997.

Dewar, John. *Dyspepsia*. New York and London, 1891.

Faithhorn, John. *Facts and Observations on Liver Complaints, and Bilious Disorders*. London, 1815.

Forth, Christopher E. 'Neurasthenia and Manhood in fin-de-siècle France.' *Cultures of Neurasthenia from Beard to the First World War*. Ed. Marijke Gijiswijt-Hofstra and Roy Porter. Amsterdam: Rodopi, 2001. 329–62.

Gaskel, Elizabeth C. *The Life of Charlotte Brontë*. Vol. 2. 2nd ed. London: Routledge, 1997.

Gibbs, Denis. 'The Demon of Dyspepsia: Some Nineteenth-Century Perceptions of Disordered Digestion.' *Gastroenterology in Britain: Historical Essays*. Ed. W.F. Bynum. London: Wellcome Institute, 1997. 29–42.

Gijiswijt-Hofstra, Marijke and Roy Porter, ed. *Cultures of Neurasthenia from Beard to the First World War*. Amsterdam: Rodopi, 2001.

Gould, George M. *Biographic Clinics: The Origin of the Ill-Health of De Quincey, Carlyle, Darwin, Huxley and Browning*. Philadelphia: Blakiston's, 1903.

Graham, Thomas John. *A Treatise on Indigestion*. 2nd ed. London, 1828.

Habershon, S.O. *On Diseases of the Stomach, the Variety of Dyspepsia*. 3rd ed. Philadelphia [London], 1879.

Harrison, Ralph. 'The Railway Journey and the Neuroses of Modernity.' *Pathologies of Travel*. Ed. Richard Wrigley and George Revill. Amsterdam: Rodopi, 2000. 229–60.

Heffer, Simon. *Moral Desperado: A Life of Thomas Carlyle*. London: Phoenix, 1995.

Holbrook, M.L. *Liver Complaint, Nervous Dyspepsia, and Headache*. New York, 1876.

Ishizuka, Hisao. 'Biographia Dyspepsia: Carlyle's Body and the Discovery of Dyspepsia.' *Shokuji no Giho* [*The Technology of Dietetics*]. Ed. Hisao Ishizuka and Suzuki Akihito. Tokyo: Keio UP, 2005. 127–46.

Jackson, Stanley W. *Melancholia and Depression: From Hippocratic Times to Modern Times*. New Haven: Yale UP, 1986.

Johnson, James. *An Essay on Morbid Sensibility of the Stomach and Bowels*. Philadelphia [London], 1827.

Kitchiner, William. *The Art of Invigorating and Prolonging Life*. 3rd ed. London, 1822.

Leared, Arthur. *The Causes and Treatment of Imperfect Digestion*. 6th ed. London, 1875.

'London, Tuesday, September 15, 1884.' Editorial. *The Times*. 16 September 1884. 15.

Oppenheim, Janet. *"Shattered Nerves": Doctors, Patients, and Depression in Victorian England*. Oxford: Oxford UP, 1991.

Porter, Roy. 'Biliousness.' *Gastroenterology in Britain: Historical Essays*. Ed. W.F. Bynum. London: Wellcome Institute, 1997. 7–28.

——. 'Nervousness, Eighteenth and Nineteenth Century Style: From Luxury to Labour.' *Cultures of Neurasthenia from Beard to the First World War*. Ed. Marijke Gijiswijt-Hofstra and Roy Porter. Amsterdam: Rodopi, 2001. 31–50.

Radkau, Joachim. 'The Neurasthenic Experience in Imperial Germany: Expeditions into Patient Records and Side-Looks upon General History.' *Cultures of Neurasthenia from Beard to the First World War*. Ed. Marijke Gijiswijt-Hofstra and Roy Porter. Amsterdam: Rodopi, 2001. 199–218.

'Review of *Medical Notes and Reflections* by Henry Holland (1839).' *Quarterly Review* 65 (1849): 315–40.

Ross, James J. 'Practical Remarks on the Treatment of the Various Forms of Dyspepsia.' *Edinburgh Medical and Surgical Review* 1.3–4 (1855): 211–20, 319–29.

Schivelbusch, Wolfgang. *The Railway Journey: The Industrialization of Time and Space in the Nineteenth Century*. Berkeley: California UP, 1986.

Sengoopta, Chandak. ' "A Mob of Incoherent Symptoms"? Neurasthenia in British Medical Discourse, 1860–1920.' *Cultures of Neurasthenia from Beard to the First World War*. Ed. Marijke Gijiswijt-Hofstra and Roy Porter. Amsterdam: Rodopi, 2001. 97–116.

Thomson, Spencer. *A Dictionary of Domestic Medicine and House Surgery*. 8th ed. London, 1859.

4
Railway Spine, Nervous Excess and the Forensic Self

Jane F. Thrailkill

> [M]emory is a material record;...the brain is scarred and seamed with infinitesimal hieroglyphics, as the features are engraved with the traces of thought and passion.
>
> Oliver Wendell Holmes, 'Mechanism in Thought and Morals', 1871

For Clifford Pyncheon in Nathaniel Hawthorne's *The House of the Seven Gables* (1851), a man who had been housed in the past, riding on a fast-moving passenger train lets loose a torrent of rapturous predictions about how modern technologies might liberate the human mind from the limitations of the material body: " 'These railroads – could but the whistle be made musical, and the rumble and the jar got rid of – are positively the greatest blessing that the ages have wrought out for us. They give us wings; they annihilate the toil and dust of pilgrimage; they spiritualize travel!' " (287) This "spiritualization", he noted, was contingent on eradicating "the rumble and the jar" of train travel, presumably because they returned one to the discomforts of the body and restrained the free play of the mind.

By contrast, a passage written 20 years later by the American scientist and novelist Oliver Wendell Holmes indicated that "some sudden jar", rather than obstructive of the mind's operations, was a critical way of making them visible: "We know very little of the contents of our minds until some sudden jar brings them to light, as an earthquake that shakes down a miser's house brings out the old stockings full of gold, and all the hoards that have hid away in holes and crannies" ('Mechanism' 282). Invoking the brain's structural fissures in his reference to a miser's "holes and crannies", the Harvard professor of anatomy focused on the mind's corporeal seat. As Holmes was writing those words, there was already a great deal of public discussion, both optimistic and anxious, over the impact of technology on the minds and bodies of inhabitants of the industrialized West. At the centre of this medical discussion was the structure of the human nervous system, which was increasingly described by research physiologists of the nineteenth century as a dispersed, information-bearing system of communication analogous to the telegraph and the railroad.

Despite the analogies between the human nervous system and the new technologies, the body was, as Shelley Trower discusses in a chapter in this book, also understood to be vulnerable to these devices for speeding communication and movement. The body's structures provided a trope for the material interconnections produced by modern technology: "[T]he whole nation", Holmes wrote in 1861,

> is now penetrated by the ramifications of a network of iron nerves [the telegraph] which flash sensation and volition backward and forward to and from towns and provinces as if they were organs and limbs of a single living body.... [T]he vast system of iron muscles [the train]...move the limbs of the mighty organism one upon another.... This perpetual intercommunication, joined to the power of instantaneous action, keeps us always alive with excitement. ('Bread' 7)

Nowhere in late nineteenth-century civilian life was the intercommunication of technological 'nerves and muscles' with their human counterparts more vividly expressed than in the consummate nineteenth-century disaster, the train wreck, with its most puzzling passenger, the mildly injured male survivor who over time developed far-flung, debilitating bodily symptoms.[1]

The figure of the hyper-symptomatic man was taken up in both fiction and medical debates of the 1880s. Oliver Wendell Holmes, for one, centred his third novel, *A Mortal Antipathy* (1885), on the baffling protagonist Maurice Kirkwood: "Everybody was trying to find out what his story was, – for a story, and a strange one, he must surely have – and nobody had succeeded" (149). This chapter examines how Holmes, in both his novel and medical writings, drew on contemporary theories about the human nervous system, reflex physiology and the body's susceptibility to shock as he negotiated the compromising diagnostic categories of insanity, effeminacy and hysteria. A person's prior experience, Holmes concluded, could be encased in the structure of the nervous system and expressed affectively through such feelings as fear.

The problem of a man whose character was radically illegible and whose symptoms exceeded initiating causes was solved – in Holmes's novel and in the medical debate over the existence of 'railway spine' – by the neurological equation of extreme emotions with physical concussion rather than congenital weakness or mental infirmity. Thus *A Mortal Antipathy* portrayed a real seat of disease, terror, in the absence of a visible physical injury. The role of the narrative, then, was to realize – in the dual sense of bringing about and bringing to consciousness – this connection: to elucidate the train of events that led from trauma to symptom, and to determine the structure of the nervous system that provided the physiological 'track' for exterior impressions to become registered in interior corporeal states.

Holmes's conception of unconscious embodied memory also provides an important pre-history for the Freudian conception of trauma, which construed traumatic experience as a primarily psychic rather than physiological wound. Yet, while the physiology of affect that emerged in the 1880s indeed saved a certain version of masculinity from hysteria understood in specifically feminine terms, it also elevated the testimony of the body, posing a challenge to traditional conceptions of willpower and rationality on which masculine character had been built.[2] Whereas the influential British physiologist Marshall Hall had in 1841 described the mind as a potentate "sitting enthroned upon the cerebrum, ... deliberating and willing, and sending forth its emissaries and plenipotentiaries, which convey its sovereign mandates, along the voluntary nerves, to muscles subdued to volition," by the decade of Holmes's novel the mind would be described, as historian Anne Harrington has recorded, as "a parliament of little men together, of whom, as also happens in real parliaments, each possesses only one single idea he is ceaselessly trying to assert" (10, 8). Holmes's narrative offered a solution to the problem of fractured corporeal authority, however: the central and centralizing role of the medical expert who, with his ethos of detachment and expertise, helped the parliament of the body to speak with one voice.

The physician, then, became an essential part of an apparatus of authentication that eschewed the incoherence, elisions and outright fabrications increasingly associated with a person's testimony and facial expressions, and instead excavated affective truths from the body's hidden depths. As one surgeon, writing in 1883, observed of nervous claimants against railway companies, "[a] perfectly steady pulse throughout the whole examination tells its own tale ... the pulse is often the only sign we have to guide us to a right estimate of a patient's condition" (159). The act of calibrating a person's conscious narrative with the unconscious testimony of his body produced a forensics of self, for it installed the use of science and technology, along with an authoritative expert, as essential for bringing to consciousness experiences pertaining to the most intimate aspects of a person's life.

Holmes, as both novelist and physician, imagined that works of literature might effectively realize these sorts of corporeal collaborations. This chapter, moreover, continues work being undertaken by contemporary scholars who argue against a reified understanding of separate spheres in the nineteenth century, for the cultural representation of nervous shock both solidified the interface between private bodies and public events and helped to produce a new way of experiencing the self in which external mediation was understood to be a way of authenticating and making available certain internal feelings.

Excess and Dissolution of the Nervous System

The problem of nervous men – of symptoms that were not just coded as feminine, such as tears and fainting spells, but that were *structurally* feminized insofar as they were in excess of generating events – was treated extensively in the discourses of medicine, and especially neurology, in the 1880s. In 1883 the physician Daniel Hack Tuke reported a case from the *Lancet* in which a "gentleman" became "much excited in connection with a very trivial occurrence": he was "seized with several paroxysms of sobbing and crying, after which he again fell into a comatose condition" (221). "The defect in this case", Tuke concluded, "lies in the absence of any apparent cause for the despondency" (221). In the words of the British physiologist Henry Maudsley in 1892, "[i]t is not natural to burst into tears because a fly settles on the forehead, as I have known a melancholic man to do" ('Suicide' 46). *A Mortal Antipathy*'s Maurice Kirkwood, with his unwillingness to speak of his past, his apparent lack of a profession and his habit of inexplicably turning pale and fainting in the presence of women, had symptoms that similarly outstripped their generating events. He appeared, in short, hysterical.

Yet, as a perplexed physician wrote to the *Lancet* on 2 January 1875, "by what figure of speech [might] an instance of deception in a man ... appropriately be termed male hysteria[?] ... The question here is, Where is the uterus?" (Hovell 37). The etymology of hysteria, *hyster* or 'uterus', marks its origin as a disease understood to be centred in a woman's reproductive organs, literally attributable to a womb that had vacated its normal location in the pelvis and 'wandered' to other parts of the body. In the winter of 1875 the *Lancet* published a series of letters to the editor that marked an uneven shift from a reproductive to a neurological understanding of hysteria. Accordingly, the physician quoted above urged that the term 'neurosis' replace hysteria, thereby acknowledging "the nervous system to be essentially the seat of the disorder" (Hovell 323). He concluded: "if a colonel of dragoons should suffer from concussion in a railway accident, I should say that he was the subject of neurosis from physical shock; but I should never be guilty of the absurdity of saying he was the subject of male hysteria, and I should prefer to say that a strong man became emotional to saying that he became hysterical" (323). Another contributor agreed: "I carefully avoid it [the term 'hysterical'] both in the wards and the lecture theatre, and am able to convey all I wish by 'emotional lesion' " (Down 108). The first doctor affirmed, "[i]t is certainly not retrograde to say that shock, and not male hysteria, is the result of a railway accident" (Hovell 323).

The repeated references to the railway accident as the exemplary shock to the masculine body are not coincidental. By mid-century, railway spine was sensationalized in lurid accounts of train wrecks featuring the figure of the travelling businessman who, though appearing physically unscathed, reaped large sums of money from insurance companies for his debilitating array of belated, mysterious symptoms. Anxiety about whether and how men might modulate the expressions of their bodies so as to maintain balance between inciting stimulus and corporeal reaction in turn raised questions about the newly discovered structure of the nervous system. The anatomists Charles Bell in England and François Magendie in France had in the first decades of the nineteenth century discerned that the nerves on the front and the back of the spinal column had different functions, one sort devoted to sensation, the other sort to movement. Marshall Hall and Johannes Muller in the 1830s elaborated on this observation by describing a 'reflex arc', in which sensory input might bypass the brain and directly stimulate muscular activity – e.g. a blink – through ganglia along the spinal column, "the twinkling of an eye being quicker than thought" (Tuke 119).

The division of the nervous system into the voluntary and involuntary systems, which were themselves further distributed into a variety of nervous centres, meant that corporeal authority was dispersed rather than centralized; as Peter Melville Logan has noted, by mid-century the "body in effect was thought to comprise many little 'brains' dispersed throughout the cerebrospinal axis, each with a type of regional authority" (166–7). In a healthy organism, the different centres were understood to be in vital equilibrium, producing concerted action and maintaining the pre-eminence of volition; as Henry Maudsley wrote in *Physiology and Pathology of the Mind* (1867), "the organization is such that a due independent local action is compatible with the proper control of a superior central authority" (54). The balance of the system, however, was delicate: if "the faculties of the spinal cord are ... exhausted by excesses of any kind," Maudsley warned, "the ill effects are manifest in degenerate action; instead of definite co-ordinate action ministering to the well-being of the individual, there ensue irregular spasmodic or convulsive movements" (71). In the schema of the neurologist John Hughlings Jackson, the highest centres located in the brain modulated the lower ones dispersed along the spinal column. Damage to the nervous system could result in what Holmes termed "the committee of the whole" ('Mechanism' 289) being, in Hughlings Jackson's words, "taken to pieces" (555). Even the nervous centres located in the brain and associated with cerebration could act independently of consciousness. As Holmes observed, "[w]e cannot always command the feelings of disgust, pity, anger, contempt, excited in us by certain presentations to our consciousness. We cannot always arrest or change the train of thoughts which is keeping us awake" ('Mechanism' 330).

Holmes, like his neurologist colleagues, repeatedly employed the figure of the train to represent the spontaneous action of the mind, referring to a train of ideas or the train of a narrative. "When we see a distant railway-train sliding by us in the same line, day after day, we infer the existence of a track which guides it" ('Mechanism' 297), Holmes pointed out in 1871. Such conceptions appeared to trouble standard accounts of agency and responsibility; a destructive drunk was "an automaton ... no more to blame for the particular acts in question than a locomotive that runs off the track is to blame for the destruction it works" ('Crime' 469). Yet the idea of the track does imply a track-layer: rather than vitiated, agency is deflected and dispersed in Holmes's telling analogy, for while the train (the equivalent of the drunk's body) might not be blamed for its derailment, the railway companies often were.

Train disasters were unnervingly frequent in the first decades of train travel, prompting a flurry of financial settlements. Historian Ralph Harrington reports that, by the early 1860s, "the railways were losing almost every personal injury case that went to court and were paying out large, and increasing, sums in compensation every year" (37). Even railway companies generally acceded to compensation for fractures, amputations and other severe physical injuries. But, as a *Lancet* editorial noted in 1861, there was

> a class of cases in which the greatest difficulty prevails. A frequent allegation is, that the plaintiff received a concussion which caused few or no serious symptoms at the time of the accident, but that a whole train of nervous symptoms, paralysis of motion or of sensation, partial or general, and impairment of the mental powers, gradually developed themselves, reaching their acme, perhaps, only after many months. (255)

In 1866, the British surgeon John Eric Erichsen published a full-length study entitled *Railway and Other Injuries of the Nervous System.*[3] Train crashes, Erichsen asserted, were uniquely modern in their catastrophic effects:

> [I]n no ordinary accident can the shock be so great as in those that occur on Railways. The rapidity of the movement, the momentum of the person injured, the suddenness of its arrest, the helplessness of the sufferers, and the natural perturbation of mind that must disturb the bravest, are all circumstances that of a necessity greatly increase the severity of the resulting injury to the nervous system, and that justly cause these cases to be considered as somewhat exceptional from ordinary accidents. (9)

These exceptional circumstances, Erichsen reported, had prompted surgeons to coin the term *"railway spine"* (9) to refer to the belated symptoms associated with survivors of rail accidents.

Erichsen's study focused exclusively on injuries where there was a "disproportion...between the apparently trifling accident that the patient has sustained, and the real and serious mischief that has occurred" (93). Granting that "it is often difficult to establish a connecting link between...[the 'mischief'] and the accident" (4), Erichsen sought to resolve the "discrepancies of opinion as to relations between apparent cause and alleged effect" (3). Erichsen argued that one could have a *functional* disorder (evident only in symptoms) without an *organic* (or lesion-based) wound: "[I]f the spine is badly jarred, shaken, or concussed by a blow or shock of any kind communicated to the body, we find that the nervous force is to a certain extent shaken out of the man, and that he has in some way lost nervous power" (95). In his early codification of the disorder, Erichsen noted that "the most remarkable phenomena" in these cases was that initially "the sufferer is usually quite unconscious that any serious accident has happened to him" (95). Once symptoms set in, however, they left the survivor "unfit for exertion and unable to attend to business" (96).

A malady prompted by the rapid pace of modern commerce, railway spine ironically transformed mobile wage earners into paralytic homebodies. Narratives of lost wages and emasculation loomed large in afflicted passengers' lawsuits against railway companies. Erichsen quoted one physician who asserted,

> 'A more melancholy object... I never beheld. The patient, naturally a handsome, middle-sized, sanguine man, of a cheerful disposition and an active mind, appeared much emaciated, stooping and dejected. He walked with a cane, but with much difficulty, and in a tottering manner.'... His saliva dribbled away; he could only utter monosyllables, and these came out, after much struggling, in a violent expiration (14).

Even Herbert Page, an outspoken critic of Erichsen's theories, describes the "uninjured" survivor of a train wreck as a pitiable sight: "Words, in fact, fail adequately to portray the distressing picture which this otherwise strong and healthy man presented" (152). Far from congenitally nervous, frail or weak, victims of railway spine were frequently described as successful professionals or burly labourers (48), underscoring the threat to masculinity broadly construed that such injuries posed.

Railway and Other Injuries of the Nervous System did not go unchallenged. Herbert Page, a railway surgeon, in 1883 published an exhaustive critique of Erichsen's conclusions, observing that litigation itself caused anxiety in train wreck survivors and noting that the litigant's relief at the conclusion of a law suit often catalyzed his physical recovery (142). Yet, as the historian Eric Caplan has remarked, this aspect of Page's rejoinder to Erichsen actually affirmed the physical reality of victims' symptoms. No longer the products of a jarred spine, a man's weeping, incontinence and even paralysis could be attributed to the emotional impact of the entire experience (from accident and diagnosis, to litigation and compensation). Page, paradoxically, opened the door to a potent possibility: that the "incidents indeed of almost every railway collision are quite sufficient – even if no bodily injury be inflicted – to produce a very serious effect upon the mind, and to be the means of bringing about a state of collapse *from fright, and from fright only*" (148, emphasis added).

The experience of fright, Page and Hughlings Jackson concurred, could come from an accident *or* from a thought or memory; the exterior and interior experiences had no firm line distinguishing them. Herbert Spencer had similarly urged, "[t]o have in a slight degree those psychical states accompanying the reception of wounds" – to hear the growls of a distant predator, say – "is to be in a state of what we call fear" (596). Daniel Hack Tuke, a physician referenced by Holmes in *A Mortal Antipathy*, succinctly summed up the corporeal equation: "a certain state of mind induces certain bodily sensations, without charging 'the subject' with imposture" (23). Indeed, Tuke concluded, "ideation, under certain circumstances, is, in its influence on the sensorium, as powerful as anything, in the outer world, which impresses the senses; and may be really more so, because in the states referred to, there is no disturbing element to distract the attention" (37). As historian Anne Harrington has observed, these nineteenth-century medical men succeeded in "neurologizing" Hume – translating the Scottish philosopher's empiricist account of human sense experience into the physiological terms of the newly authoritative field of neurology (220n1).

Embodied memory and the pathogenic secret

In 1883, the same year that Page published his comprehensive study of railway spine, the American neurologist James Jackson Putnam rehabilitated the designation 'male hysteria' – while distinguishing it from its female counterpart – to explain a set of symptoms that otherwise might be cast as fraudulent. The inability to imagine that a man might suffer from a female malady had been instrumental in Erichsen's earlier formulation of railway spine; as the surgeon had explicitly stated,

> In those cases in which a man advanced in life, of energetic business habits, of great mental activity and vigour, in no way subject to gusty fits of emotion of any kind... after the infliction of severe shock to the system, finds himself affected by a train of symptoms indicative of a serious and deep-seated injury to the nervous system – is it reasonable to say that such a man has suddenly become 'hysterical' like a love-sick girl? (qtd. in R. Harrington 52).

Putnam, in this vein, described the case of a "large, powerful and robust" labourer who seemed to heal quickly from a concussive accident (219). Months later, however, the man sought medical attention for a range of persistent symptoms, including "prostration", "wakefulness", "impairment of memory" and "emotional outbreaks" (219). Despite having "a well-marked analgesia of the entire right side", the patient "did not wince" when the neurologist scraped his back with an electrified brush (219, 220). In this way, the doctor circumvented the question of fraud by eliciting testimony directly from the patient's body.

In a crucial reformulation, Putnam suggested that, in male patients, apparently hysterical symptoms offered a corporeal record of concussive events in the past that lay outside of conscious recollection, a phenomenon that historian of science Henri Ellenberger, drawing on the fiction of Nathaniel Hawthorne, has termed the "pathogenic secret"(45).[4] Holmes built on this idea, emphasizing that 'memory' might reside in a person's body rather than his mind. The novel *A Mortal Antipathy* united the emotional aetiology of Page (that symptoms arise "from fright, and from fright only") with the corporeal aetiology of Erichsen (that symptoms attend a blow or concussion) to produce a conception of "impression" that translated a mental state such as fear or terror into a physical response by way of the autonomic nervous system: "A single impression, in a very early period of ... existence ... may establish a communication between this centre [of nervous inhibition] and the heart which will remain open ever afterwards" (*Mortal* 236). Moreover, "once the path is opened by the track of some profound impression, that same impression, if repeated, or a similar one, is likely to find the old footmarks and follow them"; this means that "the unreasoning terror of a child, of an infant, may perpetuate itself in a timidity which shames the manhood of its subject" (*Mortal* 236). Because of the physiological links between nervous and cardiac movements, Holmes wrote, "unexplained sudden deaths were of constant, of daily occurrence; that any emotion is liable to arrest the movements of life: terror, joy, good news or bad news, – anything that reaches the deeper nervous centres" (*Mortal* 222). Intellectual events, or in some cases the memory of them, could produce mortal danger.

Holmes's ideas about sudden death were likely informed by an 1884 *Lancet* article entitled 'Death of a Bridegroom'. The brief medical piece described a "recent case of sudden death", occurring a few days after the victim's wedding, attributing it to "the turbulence of his emotions" (861). The medical moral of the vignette is practically identical to that espoused by Holmes in his novel: "Surprise by joy, fright, or terror will in some cases destroy life by interrupting the normal course and performance of the vital functions.... [T]here is a risk of an untoward contingency when the nervous system is so agitated by strong excitation of any particular centre as to disturb the harmony of its working" (861).

The centrality of emotion to vital functioning subtended a new account of memory in the nineteenth century. Events from the past were not simply stored in the brain; instead, memory was understood to reanimate parts of the body that had been originally affected. The psychologist Théodule Ribot called this process "organic memory", which he described as "the acquired movements which constitute the memory of different organs – the eyes, hands, arm, and legs.... A rich and extensive memory is not a collection of impressions, but an accumulation of dynamical associations" (31). Ribot described how even thinking can function in this way: "Unconscious cerebration does its work noiselessly, and sets obscure ideas in order. In a curious case related by Dr. Holmes and cited by [W. Benjamin] Carpenter, a man had a vague knowledge of the work going on in his brain, without attaining to the state of distinct consciousness" (37). For the nineteenth-century thinkers, the structure of the nervous system facilitated an account whereby a series of apparent oxymorons could become explicable: memory without memory, feeling without an object, and indeed thinking without consciousness.

In Holmes's *A Mortal Antipathy*, Maurice's autobiographical narrative finally accounts for his seemingly excessive symptoms. Maurice recounts to the local physician Dr. Butts how he fell from a nurse's arms as a baby, the shock thereby transforming a "pleasant, smiling infant, with nothing to indicate any peculiar nervous susceptibility" (207) into a man who faints in the presence of women. Pieced together from "circumstances as told me and vaguely remembered" (207), Maurice's account echoes the "paralyzing terror" of train-wreck survivors:

> That dreadful experience is burned deep into my memory. The sudden apparition of the girl; the sense of being torn away from the protecting arms around me; the frantic effort to escape; the shriek that accompanied my fall through what must have seemed unmeasurable space; the cruel lacerations of the piercing and rending thorns, – all these fearful impressions blended into one paralyzing terror. (209)

Maurice's "memory", corporeally registered by the terror, is nonlinguistic for he "had no thought, living like other infants the life of impressions without language to connect them in a series" (224).

Here we begin to see important distinctions between Holmes and the theorists of railway spine. In describing Maurice's 'memory' as existing *in the absence of* consciousness, Holmes stripped the pathogenic secret of its painful location in the consciousness of the subject, while the "memory" of it, no longer understood to be symbolic of a prior event or the physical concomitant to a conflict of conscience, was affirmed to be *commensurate with its physical manifestations*. Memory, Ribot maintained, was "a biological fact: it is an impregnation" (196) of experience onto the muscles, tissues, sensory organs and – frequently but not necessarily – the brain. The operation of this new sort of memory is vividly expressed in *A Mortal Antipathy*, when Maurice goes into a state of shock in the presence of women:

> The cause of this violent and appalling seizure was but too obvious. The approach of the young girl and the dread that she was about to lay her hand upon me had called up the same train of effects which the moment of terror and pain had already occasioned.... It was too evident that a chain of nervous disturbances had been set up in my system which repeated itself whenever the original impression gave the first impulse. (210–11)

The term *impression* could of course imply two different functions: the processing of a mental image thereby involving something akin to conscious memory, or the physiological inscription (pressing in) of experience directly onto the nervous system. That Holmes meant the latter is indicated in a scene where Maurice inexplicably experiences a "change of color, anxiety about the region of the heart, and sudden failure as if about to fall into a deadly fainting-fit" (217) – only to find that there was a woman sitting nearby but out of his line of vision. A rather silly scene, it nonetheless serves the purpose of producing a situation that cannot be explained away as a 'mere' mental event, or even some odd fabrication.

The forensic self

Maurice, in both inhabiting his corporeal experience and possessing a narrative of explanation, "had learned to look upon himself very much as he would upon an intimate *not* himself, – upon a different personality" (229). Holmes's character exemplifies what psychiatrist Pierre Janet in 1891 termed the "undoubling" (*dédoublement*) of a personality:

> all the psychological phenomena that are produced in the brain are not brought together in one and the same personal perception; a portion remains independent under the form of sensations or elementary images, or else is grouped more or less completely and tends

to form a new system, a personality independent of the first. These two personalities are not content merely to alternate, to succeed each other; they can coexist in a way more or less complete. (*Mental* 492–3)

As historian Ruth Leys has noted, modern theorists of trauma have discerned in Janet's writings a distinction between " 'traumatic memory,' which merely and unconsciously *repeats* the past, and 'narrative memory,' which *narrates the past as past*"; Janet, moreover, "validat[ed] the idea that the goal of therapy is to convert 'traumatic memory' into 'narrative memory' by getting the patient to recount his or her history" (105). Maurice's lack of conscious memory, Holmes was at pains to assert, inhered in his being a baby when his shocking fall occurred. (This is opposed to Freud's dynamic unconscious, in which ideas or desires found socially unacceptable are pushed out of the realm of consciousness.) Moreover, Maurice's emotional recounting of his infantile trauma – pieced together journalistically from eyewitness accounts and medical explanations – does not alleviate his symptoms.

Holmes in the novel focused his discussion of memory not on narratives of cases, but on the brain and the nervous system. Maurice's problems are, finally, not psychological but 'telegraphic', i.e. neurological:

The brain, as all know, is the seat of ideas, emotions, volition. It is the great central telegraphic station with which many lesser centres are in close relations, from which they receive, and to which they transmit, their messages. The heart has its own little brains, so to speak, – small collections of nervous substance which govern its rhythmical motions under ordinary conditions. … There are two among the special groups of nerve-cells which produce directly opposite effects. One of these has the power of accelerating the action of the heart, while the other has the power of retarding, or arresting this action. (*Mortal* 235)

In Holmes's explanation, the communication between the two systems was mediated by emotion. Thus, "[a] single impression, in a very early period" of a person's life, established "a communication between this centre and the heart which will remain open ever afterwards"; and, as with any repeated motion, "[h]abit only makes the path easier to traverse" (*Mortal* 236).

So while a rush of feeling might overload the nervous system, Daniel Hack Tuke, thinking homeopathically, hypothesized that "[f]ear may heal as well as cause disease" (333). The plot of *A Mortal Antipathy* produces just such a dose of therapeutic panic, when a fire breaks out in Maurice's house:

> The dread moment which had blighted his life returned in all its terror. He felt the convulsive spring in the form of a faint, impotent spasm, – the rush of air, – the thorns of the stinging and lacerating cradle into which he was precipitated. One after another those paralyzing seizures which had been like deadening blows on the naked heart seemed to repeat themselves, as real as at the moment of their occurrence. (265)

Rather than a memory producing feelings, Maurice experiences the reverse: emotion catalyzes the memory. Holmes invokes the technology of the photograph to figure the corporeal inscription of memory: "The sensitive plate [the body] has taken one look at the scene, and remembers it all. Every little circumstance is there ... but invisible; potentially present, but impalpable, inappreciable, as if not existing at all" (265). Feelings provide the "wash" to make the past, housed in the body's dynamic structures, visible: "the rush of unwonted emotion floods the undeveloped pictures of vanished years, stored away in the memory ... and in one swift instant the past comes out as vividly as if it were again the present" (265–6). It is in this moment of terror that the past and present converge in real time, and it is Euthymia 'The Wonder' (rather than Lurida, whose nickname is 'The Terror') who rescues Maurice from the burning building and effects "the revolution in his nervous system which would be the beginning of a new existence" (277).

By relocating the terms of the debate over railway spine from adulthood to childhood, and from biography to neurology, Holmes established a non-pejorative explanation for hyper-symptomatic men. The neurological track of fear, developed at an early, somatically pliable period, could be 'greased' through habit, catalyzing panic from increasingly insignificant events (e.g. a fly settling on the forehead). Beyond the volition or indeed consciousness of the individual, the path of transmission was conceived as corporeal and dynamic. Twentieth- and twenty-first century trauma theories have tended to focus on the integration of past traumatic memories (their translation into narrative memory). Holmes, by contrast, affirmed the need for *both* a narrative remembering *and* a corporeal 'forgetting'. Ruth Leys has highlighted Pierre Janet's insistence that "narrated recollection was insufficient for the cure. A supplementary action was required, one that involved a process of 'liquidation' that, terminologically, sounded suspiciously like 'exorcism' or forgetting" (105). As Janet wrote in 1923 about one of his female patients, "[t]o forget the past is in reality to change behavior in the present. When she achieves this new behavior, it matters little whether she still retains the verbal memory of her adventure, she is cured of her neuropathological disorders" (*Médicine* 126).

So while Leys and other historians of trauma have rightly emphasized the importance of Janet's part in a genealogy of trauma and dissociation, the work of Oliver Wendell Holmes should be similarly recognized, as an influential US articulation of a distinctly modern form of selfhood.[5] Double consciousness, as articulated in Holmes's literary and scientific writing, helped to install narrative – more precisely, the story of an individual's life experiences – in an absolutely central yet crucially vexed position vis-à-vis the subject. For while narrative is a crucial technology for a form of subjectivity that requires the unearthing and then uprooting of prior pathogenic experiences to produce an aetiology (or explanatory 'track') of bodily symptoms, it is also that which may be unavailable to the conscious mind of the person in possession of the memories, who relives rather than represents the past.

The legacy of this way of conceiving of the self, then, is that corporeal symptoms are increasingly understood to be inscriptions of prior experiences and therefore, unlike narrative, to provide unmediated access to a person's past. Holmes's Dr. Butts listens to Maurice's story, but he also keeps his fingers on the young man's wrist, calibrating his tale with the testimony of his body. This forensic model underwrites the role of a detached expert who, in tracing effects back to precipitating causes in a person's life, employs narrative as a crucial diagnostic tool. Narrative, in other words, is at once repudiated and installed as central to a physiological psychology. Here we encounter the enabling conditions for literary realism and its practitioners, who hover uncomfortably between fabricating events and mobilizing an apparatus of expertise to elicit the innermost secrets of persons and society. It is my contention that, in the person of Holmes, at least, novelists were indeed instrumental in helping realize (in Ian Hacking's sense of making up) a new relationship between a person and his or her past, a whole new way of conceiving of oneself as a person in the nineteenth century. Holmes succeeded in establishing a masculinity that was vexed, at best: he saved his protagonist from excessiveness at the expense of a radically refigured autonomy, produced narrative resolution at the cost of a unified self, and helped to establish the terror-stricken man, rather than the hysterical woman, as the elided foundational figure for modern trauma theory.

Notes

1. See Jessica Meyer's chapter in this collection for an example of the place of this aetiology in producing the term 'shell shock' in 1915.
2. Though we concur about the equation of emotional and physical shock in Holmes's last novel and in the discourse of railway spine, Randall Knoper emphasizes Holmes's linkage of neurological damage to sexual 'inversion'. In his compelling essay, Knoper positions Holmes's novel within "a history of the medicalization of homosexuality" (14). For a useful contextualization of scientific themes Holmes's three novels (*Elsie Venner*, *The Guardian Angel* and *A Mortal Antipathy*), see Boewe, 'Reflex Action'. More recently, Peter Gibian has framed Holmes' self-titled 'medicated novels' in the context of discourse and dialogue, casting the narrative voice as that of the "doctor-confessor-psychoanalyst hero" (71).
3. I am grateful to Karen M. Odden for bringing to my attention the British literature on railway spine, in her unpublished essay 'Problems with Railways, Problems with Stories: A Narrative History of the Origins of Trauma, 1840–1890'.
4. "In 1850," the French historian writes, "Nathaniel Hawthorne described, in his masterpiece *The Scarlet Letter*, how a pathogenic secret can be discovered by a wicked man and exploited in order to torture his victim to death" (45).
5. See also Young, *Harmony*, Antze and Lambek, *Tense Past*, and Hacking, *Rewriting*.

Works cited

Antze, Paul and Michael Lambek, eds. *Tense Past: Cultural Essays in Trauma and Memory*. New York: Routledge, 1996.

Boewe, Charles. 'Reflex Action in the Novels of Oliver Wendell Holmes.' *American Literature* 26.3 (November 1954): 303–19.

Caplan, Eric. *Mind Games: American Culture and the Birth of Psychotherapy*. Berkeley: California UP, 2001.

'Death of a Bridegroom.' *Lancet*. 10 May 1884. 861.

Down, J. Langdon. Letter to the editor. *Lancet*. 16 June 1875. 108.

Editorial. *Lancet*. 14 September 1861. 255–56.

Ellenberger, Henri. *The Discovery of the Unconscious: The History and Evolution of Dynamic Psychiatry*. New York: Basic Books, 1970.

Erichsen, John Eric. *On Railway and Other Injuries of the Nervous System*. London: Walton & Maberly, 1866.

Gibian, Peter. *Oliver Wendell Holmes and the Culture of Conversation*. Cambridge: Cambridge UP, 2001.

Hacking, Ian. *Rewriting the Soul: Multiple Personality and the Sciences of Memory*. Princeton, N.J.: Princeton UP, 1995.

Harrington, Anne. *Medicine, Mind, and the Double Brain: A Study in Nineteenth-Century Thought*. Princeton, N. J.: Princeton UP, 1989.

Harrington, Ralph. 'The Railway Accident: Trains, Trauma, and Technological Crises in Nineteenth-Century Britain.' *Traumatic Pasts: History, Psychiatry, and Trauma in the Modern Age, 1870–1930*. Ed. Mark S. Micale and Paul Lerner. Cambridge: Cambridge UP, 2001.

Hawthorne, Nathaniel. *The House of the Seven Gables.* 1851. New York: Oxford UP, 1991.

Holmes, Oliver Wendell. 'Bread and the Newspaper.' *The Works of Oliver Wendell Holmes.* Vol. 8. Boston: Houghton Mifflin, 1892.

——. 'Crime and Automatism.' *The Atlantic Monthly* 35.210 (April 1875): 466–82.

——. 'Mechanism in Thought and Morals.' 1871. *Pages from an Old Volume of Life: A Collection of Essays, 1857–1881. The Works of Oliver Wendell Holmes.* Vol. 8. Boston: Houghton Mifflin, 1892.

Hovell, D. De Berdt. 'Male Hysteria (?).' Letter to the editor. *Lancet* (2 January 1875): 37.

——. 'Not Hysteria, but Neurosis.' Letter to the editor. *Lancet* (27 February 1875): 323.

Hughlings Jackson, John. 'Evolution and Dissolution of the Nervous System.' *Lancet* (12 April 1884): 554–61.

Janet, Pierre. *The Mental State of Hystericals: A Study of Mental Stigmata and Mental Accidents.* Trans. Caroline Rollin Corson. 1901. Washington, D.C.: University Publications of America, 1977.

——. *La Médicine Psychologique.* 1923. Paris, 1980.

Knoper, Randall. 'Trauma and Sexual Inversion, circa 1885: Dr. Holmes's *A Mortal Antipathy* and Maladies of Representation.' *Neurology and Literature, 1860–1920.* Ed. Anne Stiles. Basingstoke: Palgrave, 2007. 119–40.

Leys, Ruth. *Trauma: A Genealogy.* Chicago: Chicago UP, 2000.

Logan, Peter Melville. *Nerves and Narrative: A Cultural History of Hysteria in Nineteenth-Century British Prose.* Berkeley: California UP, 1997.

Maudsley, Henry. *Physiology and Pathology of the Mind* [1867]. Washington, D.C.: University Publications of America, 1977.

——. 'Suicide in Simple Melancholy.' *Medical Magazine* 1 (1892–3).

Page, Herbert. *Injuries of the Spine and Spinal Cord without Apparent Mechanical Lesion and Nervous Shock in Their Surgical and Medico-Legal Aspect.* London: J. & A. Churchill, 1883.

Putnam, James Jackson. 'Recent Investigations into the Pathology of So-Called Concussion of the Spine.' *Boston Medical and Surgical Journal* 109.10 (6 September 1883): 217–20.

Ribot, Théodule. *Diseases of Memory: An Essay in the Positive Psychology.* Trans. William Hungtinton Smith. New York: Appleton, 1882.

Spencer, Herbert. *The Principles of Psychology.* London: Longman, Brown, Green & Longmans, 1855.

Tuke, Daniel Hack. *Illustrations of the Influence of the Mind upon the Body in Health and Disease, Designed to Elucidate the Action of the Imagination.* Philadelphia: Henry C. Lea, 1873.

Young, Alan. *The Harmony of Illusions: Inventing Post-Traumatic Stress Disorder.* Princeton, N.J.: Princeton UP, 1995.

5
"The Conviction of its Existence": Silas Weir Mitchell, Phantom Limbs and Phantom Bodies in Neurology and Spiritualism

Aura Satz

In 1866, the American neurologist Silas Weir Mitchell anonymously (and, in his account, unwittingly) published a fictional account entitled 'The Case of George Dedlow' in the *Atlantic Monthly*.[1] Dedlow, the narrator of the story, describes, in the first person, how during the Civil War he is gradually reduced to a "useless torso, more like some strange larval creature than anything of human shape" (129), having had both his arms and legs amputated. Increasingly throughout the second half of the nineteenth century, amputees became a highly visible part of the population, due to both the accidents of the industrialization of modernity and, in America, the devastating effects of the Civil War of 1861–5. This visibly dismembered citizen corresponded also to a new understanding of the body in neurological terms, as a series of imperceptible symptoms began to haunt their bodies, demanding to be taken seriously as medical occurrences invisible to the human eye. In the story, George Dedlow, himself a doctor (clearly ghost-writing Mitchell's medical experiences as a physician on the battlefield), objectively describes phantom limb phenomena of his own body and of other fellow amputees. In 1871, five years after 'The Case of George Dedlow', Mitchell published a neurological account in an article entitled 'Phantom limbs', which appeared in *Lippincott's Magazine of Popular Literature and Science*, thus saving the phenomena from centuries of medical eclipse. Mitchell was to become known as the 'father of modern neurology' in America, but was also an accomplished novelist in his own right, especially from the 1880s onwards. 'The Case of George Dedlow' was the only occasion in which he published his medical discoveries in fictional form prior to publishing them as scientific literature. Fascinatingly, the story ends with a spiritualist séance during which Dedlow's missing legs are called forth. To his sceptical surprise, the limbs appear and he is momentarily re-embodied. The story was supposedly intended as a parody of spiritualist belief,[2] yet it was immensely popular, so much so that as a result it solicited donations to the Philadelphia Stump Hospital where it was set, and was endorsed by spiritualists as proof of the afterlife. It seems the story was received as evidence not only of the conviction of the existence of phantom limbs, but of phantoms more generally. Through the fictional narratives and scientific texts of Mitchell, this chapter will look at the surprising parallels between the study of phantom limb phenomena and the movement of spiritualism, and how the two intertwine in their attempts to map invisible forces, employing at times a similar rhetoric of substantiation.

Phantom limb phenomena had been observed since the Middle Ages, although for the most part they were recounted in the context of miraculous limb restoration (cf. Price and Twombly), shrouded in the imaginative veils of folkloric narrative. Even Ambroise Paré, the sixteenth-century military surgeon who first addressed the phenomena in medical terms, described pain in the amputated part as "a thing worthy of wonder and almost incredible" (221). Hence it comes as little surprise that Mitchell chose fiction as the first guise under which to present his discoveries concerning phantom limb phenomena. But rather than render the phenomenon as a miraculous event perceived from the outside, he chose the empathetic viewpoint of the person experiencing these phantoms, through whom he conveyed some of his medical observations as a surgeon during the American civil war (Dedlow's character is simultaneously a diagnosing doctor and a suffering patient). But he could not resist the lure of something supernatural, and indeed spiritualism serves as the mysterious framework through which to describe the remarkable sensory hallucination of the experience of a phantom limb. Mitchell actually coined the term 'phantom limb' in his 1871 article, and by doing so he was clearly tapping into this spectral phantasmagoric vogue which prevailed during his life-time.[3] The term 'phantom' was first used in the English language as *fantum* around 1300, meaning 'illusion, unreality'. The meaning as 'specter, spirit, ghost' is attested from 1382; that of 'something having the form or appearance, but not the substance, of some other thing' is from 1707.[4] Phantoms thus imply an inherent tension between supernatural phenomena and psychic or mental phenomena, which was intensified by the absorption of the term 'phantasmagoria' into the "spectralization or 'ghostifying' of mental space" (Castle 141–2). That Mitchell should have chosen this terminology undoubtedly alluded to the prevalent cultural force that was spiritualism during the mid-nineteenth century. Before going into the depths of Mitchell's curious narrative, it is worth giving some history of spiritualism.

The movement of spiritualism was allegedly born on 31 March 1848, when two young girls aged 11 and 15, Kate and Margaretta Fox, made a breakthrough by contacting the entity who had been disturbing the family house in Hydesville, upstate New York, with strange sounds and activities. The Fox Sisters were later joined by their elder sister Leah, and they began to tour and exhibit their mediumship skills. In 1850 they founded spiritualism as a church, and by 1860 the federal census listed 17 spiritualist churches, which by 1906 had grown to 455. National organization efforts began in the 1860s, and the National Spiritualist Association of Churches was founded in Chicago in 1893. The movement found enthusiasts in Britain and France, and was soon a worldwide trend, however disorganized and unregulated. The basic tenet of spiritualism were that the dead communicate with the living, providing empirical proof of the immortality of the soul, and indeed it flourished as a result of the fatalities caused by the American Civil War and the First World War. But perhaps the movement's popularity was due to the broader implications and possibilities enabled by spirit/ghost-writing, mediumship and trance-lecturers. This removed mode of authorship, which enabled the medium to be a mouth-piece for the words of others, was taken up predominantly by women, abolitionists and dissenters, as a mode of un-inhibited, non-institutionalized, un-hierarchical expression. The movement was extremely democratic in that it had no orthodox doctrine, no official leadership, and mediums received no training and no ordination. Often séances took place in the privacy of the domestic setting, around the kitchen table. Indeed, the movement drew from the reformism of the time, and many of the early participants in spiritualism were radical Quakers and other reformers intent on promoting temperance, fighting slavery and advancing women's rights. Thus spiritualists could communicate not only the messages of the past, but also more pressing ideological themes of the present which were lacking an appropriate platform. That said, as Daniel Cottom points out, spiritualism highlighted the problem of communication, as often the messages conveyed by the spirits were misspelled, uneventful and presented a rather undignified version of the supernatural, so that ultimately spiritualism "made communication more important, [whilst] it made interpretation less certain" (56).

Spiritualism was also significant in that it incorporated into its rather chaotic practice the new technologies of modernity. So the inventions of the telegraph, developed from the 1830s onwards, and at the end of the nineteenth century wireless telegraphy or radio, coincide with the morse-code rappings and disembodied voices of the spiritualists (one of the movement's main newspapers was called *The Celestial Telegraph*), not to mention the invention of photography in 1839, which likewise facilitated the capturing of countless spirits in ghostly mirage form. In 1876 and 1877 the telephone and gramophone were invented, and incorporated into spiritualist language. All of these media, which so vividly preserved the voices and images of the absent, were used as means of producing substantiating evidence of the existence of life after death, as well as serving as useful utopian metaphors to describe the technicalities of how the departed might make contact or materialize (Sconce 25). Spiritualism strove to appropriate some of the prestige of empirical science, and the new technologies at its disposal. Concurrent with the ever-expanding practices of spiritualism through the proliferation of mediums and séances, investigation committees were set up to prove the fraudulent activities of supposed humbugs and impostors. Spiritualism indeed became a profitable spectacle, and alongside demonstrations of mediumship and séances, debunking performances of the very same acts were conducted by disbelieving magicians out to expose the hoax (Harry Houdini would be among the most famous of these anti-spiritualist debunkers).

Though seemingly incompatible, spiritualism and neuroscience did at times cross paths, not least in their timeline. In America, neurology established both a national journal and a national organization during the 1870s, simultaneous with the organizational efforts of spiritualism. Neurologists turned their attention to spiritualist phenomena such as trance and somnambulism, and frequently construed mediumship as a neurological disorder. Neuroscience was obviously propelled into uncharted discoveries thanks to the new technologies, but perhaps more compellingly one could say the new technologies were themselves neurologically inclined, as they extended the central nervous system (the telegraph as an artificial mouth, the telephone an artificial ear), and so the gramophone could serve as a timely illustration of how the brain works and stores information (Kittler 28). And just as spiritualism was offering an alternative to scopic authority, whereby knowledge was gleaned through witnessing phenomena rather than seeing them (Weinstein 127), neuroscience, in its ambition to actually see brain function (only to be fulfilled about a century later), was forced in the meanwhile to articulate new modes of non-scopic inference and diagnosis.[5]

Mitchell chose spiritualism as the fictive construct which enabled the re-integration of the unhappy "fraction of a man" (149) that Dedlow had become after having lost 80 percent of his bodily mass in a series of amputations. By the date of the story's publication, one year after the end of the Civil War, spiritualism had become a vital national movement. His detailed account implies that he must have attended at least one of these events, if nothing else than to serve as source material for his story. It is illuminating that Mitchell – though allegedly a rational doctor satirizing spiritualism and its practitioners – also hints at less sceptical inclinations, conveying the movement in a rather compassionate light. The sergeant who invites Dedlow to the séance claims the New Church is "a great comfort" for a plain man like himself when he is weary and sick, to which Dedlow replies "if only one could believe it" (141–2). Spiritualist faith features as consolation to the simple folk traumatized by the losses of war, and Dedlow's initial incredulity was the position taken by many who desperately sought such consolation but could not rid themselves of the prejudice against such populist, vulgar and tasteless beliefs. But disbelief was not a drawback. As Ann Braude writes, initially spiritualism required that its followers believe nothing; rather, it asked that they become "investigators" to observe "demonstrations" under "test conditions" of the séance room (4). It did not presuppose conversion, but along the lines of the Doubting Thomas paradigm, offered the path of conversion through tangible evidence. And so Dedlow agrees to attend the sitting, though describing with some condescension the people who would have gone to these circles: a flabby "eclectic doctor, who had tried his hand at medicine and several of its quackish variations, finally settling down on eclecticism...a female-authoress, I think, of two somewhat feeble novels...a pallid, care-worn young woman...a magnetic patient of the doctor." Others, like Dedlow, he describes as "strangers brought hither by mere curiosity" (143–4). Mitchell portrays the medium as a greedy con artist, who keenly scans the faces of his applicants whilst spelling out the names of the dead with whom they wish to communicate. Frequently cynics attended the séances to try and debunk the fraudulent medium, and indeed Mitchell describes one such attendee: "a stolid personage of disbelieving type, [with whom] every attempt failed, until at last the spirits signified by knocks that he was a disturbing agency, and that while he remained all our efforts would fail." And so he leaves "with a skeptical sneer at the whole performance" (146).

Such sifting of true manifestations of invisible forces from cases of deception was also a significant concern of Mitchell's within the medical profession. In 1864 he circulated a pamphlet 'On Malingering', the faking of illness, whereby the patient, and, on occasions, the doctor too, was put to trial. The pamphlet outlined various ways in which the physician should rather pitilessly prod, poke, cut and more generally survey the patient to expose the malingerer of nervous disease. Not unlike the debunkers intent on trying spiritualist phenomena under increasingly scientific apparatuses and modes of measuring, quantifying and validating that bordered on the torturous, the doctor became a prosecutor of sorts, equipped not only with medical knowledge but also, to quote Mitchell, with "ingenuity" (371). In these early stages of mapping out the invisible, often organically undetectable symptoms of nervous illness, malingerers were the bane of the empirical doctors of the mid-nineteenth century, as they "purposely blurred the boundaries of illness, health, and self-representation that were already so fuzzy in medical discourse of the period" (Long 44). But Mitchell's suspicion does not stop at the patient, for doctors could also require examination, as they too might be desperately seeking means of substantiating their ground-breaking diagnosis of some invisible nerve disease and thus maintain control of the hospital. In Mitchell's short story *Autobiography of a Quack*, published in book-form in 1900 together with 'The Case of George Dedlow', he turns against his own line of occupation, so to speak, by caricaturing an ambitiously corrupt doctor of "large and liberal views" in his profession, who decides to combine the various "irregular" practices of medicine, homeopathy, spiritualism and "all the available isms" for sheer profit (73).

The slipperiness of objective knowledge when so many illnesses were knowable only through the narratives of subjective experience gave leeway to misdiagnosis, as well as fraudulent diagnosis, and naturally the diagnosis of all things psychogenic (as it was termed in 1902), originating in the mind rather than in the physiological body. Let us not forget that neurology is to an extent the forefather of psychoanalysis, as Freud himself had been a neurologist in his youth, studying cases of hysteria under Charcot and writing papers on aphasia and other neurological disorders.[6] Interestingly, Freud is credited with coining the word 'agnosia' in his 1891 book *On Aphasia* (78), from the Greek *agnostos* (unknown, unknowable), meaning 'the loss of the ability to recognize people or objects due to brain damage' – an indicative coincidence as so much of the medical profession around this time was based on a process of diagnosis of the invisible, the impenetrable and the elusive.

The quest for knowledge of the indiscernible meant that other forms of acumen came to the fore, particularly verbal accounts. The reliance on narrative, both of the patients' descriptions of their imperceptible symptoms, and of the observations of the doctor, foreshadows the techniques of Freudian psychoanalysis.[7] That said, Mitchell himself subscribed to a distinctly physiological perspective, and did not acknowledge the centrality of his patients' storytelling. Indeed, in 1908 he wrote an article in the *Journal of the American Medical Association*, where he mocked those "psychopaths" who believed "the chief value of [his] treatment lay in its psychotherapy" ('The Treatment by Rest' 2033),[8] and even claimed that in some simple medical cases "intrusive psychotherapy" can do more harm than good (2035).

Mitchell played a key role in this process of archiving previously unmapped symptoms and experiences caused by the post-bellum damages of the Civil War, and his 1864 book *Gunshot Wounds and Injuries of Nerves* described, in his words, a collection of phenomena "of suffering as yet undescribed" (101). The neurologically injured bodies of the Civil War were alienated, detached, out of synch with their normal functions, often experiencing pain 'incorrectly' as it were. Some felt pain in the wrong place; others did not feel pain at all when they should have. One soldier who had been shot in the left cheek found that he could not use his left arm, which even later remained weak (45); another was shot near the left eye through to the jaw, and never felt pain, "rarely more than sore" (*Injuries of Nerves* 327). Mitchell wrote of the difficulty of diagnosis: "I have, indeed, great doubt as to our capacity to distinguish the form of injury from the resultant symptoms alone" (222). In other words, the visible wound and the interior pain did not necessarily correspond. The body appeared to be estranged from itself, experiencing a wide range of sensory hallucinations of presence where there should be absence and vice versa. Phantom limb pain is invisible from the outside, and in his medical descriptions Mitchell performs an "exquisite rhetorical sleight of hand" (Herschbach 192), transforming patients' stories of invisible sensations into facts, knowable objects of nosological description, only very rarely embellishing them with 'the patient describes', 'he says' or an actual citation. Pain objectively *is*, rather than subjectively *feels like* (this is most palpable in *Gunshot Wounds*). In his 1872 book *Injuries of Nerves*, the final section on "Neural Maladies of Stumps" (343–68) focuses on sensory hallucinations, and here too the phantom limb seamlessly alternates between being, appearing, seeming and feeling like. Likewise, in the story Mitchell oscillates between phenomenologically materializing Dedlow's imperceptible phantom himself, and observing it as an outsider. Dedlow, at the point in the story of being a quadruple amputee, amuses himself by learning from other 'limbless folk', and from his own experience, the peculiar feelings associated with lost limbs:

> I found that the great mass of men who had undergone amputations for many months felt the usual consciousness that they still had the lost limb. It itched or pained, or was cramped, but never felt hot or cold. If they had painful sensations referred to it, *the conviction of its existence* continued unaltered for long periods; but where no pain was felt in it, then by degrees the sense of having that limb faded away entirely.... Where the pains come and go, as they do in certain cases, the subjective sensations thus occasioned are very curious, since in such cases the man loses and gains, and loses and regains, the consciousness of the presence of the lost parts, so that he will tell you, "Now I feel my thumb, now I feel my little finger." I should also add that nearly every person who has lost an arm above the elbow feels as though the lost member were bent at the elbow, and at times is vividly impressed with the notion that his fingers are strongly flexed.
>
> Other persons present a peculiarity which I am at a loss to account for. Where the leg, for instance, has been lost, they feel as if the foot were present, but as though the leg were shortened. Thus, if the thigh has been taken off, there seems to them to be a foot at the knee; if the arm, a hand seems to be at the elbow, or attached to the stump itself. (131–3, emphasis added)

Such accounts of metamorphic experiences, telescoping limbs and bodily phantasmagoria, reveal the instability not only of the body but also of its proper diagnosis. This is a body that is to some extent inarticulate and inarticulable, suffering from agnosia, a loss of knowledge, and often even anosognosia (coined by neurologist Joseph Babinski in 1914) – the brain's impaired awareness (or even denial) of disability.

Phantom limb phenomena, perhaps more than any other nervous symptom, question the very sense of self, identity and ownership over one's self. As Dedlow speculates:

> Thus one half of me was absent or functionally dead. This set me to thinking how much a man might lose and yet live.... Would such a being, I asked myself, possess the sense of individuality in its usual completeness, even if his organs of sensation remained, and he were capable of consciousness? Of course, without them, he could not

have it any more than a dahlia or a tulip. But with them – how then? I concluded that it would be at a minimum, and that, if utter loss of relation to the outer world were capable of destroying a man's consciousness of himself, the destruction of half of his sensitive surfaces might well occasion, in a less degree, a like result, and so diminish his sense of individual existence. I thus reached the conclusion that a man is not his brain, or any one part of it, but all of his economy, and that to lose any part must lessen this sense of his own existence. (138–9)

By the time George Dedlow has lost all of his limbs he is not only struggling to preserve his identity, but also his humanity. The possession of his sense of self is in doubt. In Mitchell's article, supposedly his scientific rectification of the story, he writes that "so far as we are aware, no one survived the removal of all four limbs above the elbows and knees, although such a case is said to have occurred in the Napoleonic wars" ('Phantom Limbs' 564). Dedlow is therefore already a medical anomaly, a man wavering on the threshold of life and death, having only just survived amputation of all limbs including the thighs "very high up" (127). This unhappy "fraction of a man" (149) is himself already a ghost of sorts, barely alive, living in a constant purgatory, his hand "dead except to pain" – to which a visitor retorts, "such will you be if you die in your sins: you will go where only pain can be felt. For all eternity, all of you will be just like that hand – knowing pain only" (122). Whilst Dedlow is haunted and perplexed by this sense of diminished existence, his attention is caught by the words of a spiritualist, a sergeant who described his belief in the afterlife thus:

["]should not the dead soul talk to the living? In space, no doubt, exist all forms of matter, merely in finer, more ethereal being. You can't suppose a naked soul moving about without a bodily garment – no creed teaches that; and if its new clothing be of like substance to ours, only of ethereal fineness, – a more delicate recrystallization about the eternal spiritual nucleus, – must it not then possess powers as much more delicate and refined as is the new material in which it is reclad?" "Not very clear," I answered; "but, after all, the thing should be susceptible of some form of proof to our present senses." (142)

Phantom limbs are indeed susceptible to "some form of proof to our present senses". Mitchell indirectly points to parallels between the afterlife of the amputated limb and the afterlife of a dead soul. This analogy is also made in his later 1871 article on 'Phantom Limbs' (occasionally replaced by the synonym 'spirit member'), where he describes a case of a man who for two years had ceased to be conscious of the limb. When he was given an electrical current, ignorant of its possible effects, "he started up, crying aloud, 'Oh, the hand, the hand!' and tried to seize it with the living grasp of sound fingers. No resurrection of the dead, no answer of a summoned spirit, could have been more startling" (566). Mitchell was reworking this case from his 1864 book, where the comparison with spiritualism led him to write himself, the doctor, into the role of conjuring medium: "The phantom I had conjured up swiftly disappeared, but no spirit could have more amazed the man, so real did it seem" (*Injuries of Nerves* 349).

In one of his most evocative passages, Mitchell touchingly describes the condition of a person experiencing a phantom limb as being

> haunted, as it were, by a constant or inconstant fractional phantom of so much of himself as was lopped away – an unseen ghost of the lost part, and sometimes a presence made sorely inconvenient by the fact that while but faintly felt at times, it is at others acutely called to his attention by the pains or irritations which it appears to suffer from a blow on the stump or a change in the weather.
>
> There is something almost tragical [sic], something ghastly, in the notion of these thousands of spirit limbs haunting as many good soldiers, and every now and then tormenting them with the disappointments which arise when, the memory being off guard for a moment, the keen sense of the limb's presence betrays the man into some effort, the failure of which of a sudden reminds him of his loss. ('Phantom Limbs' 565–6)

The loss of a limb becomes analogous to the loss of a person, haunting the survivors and tormenting them with the inconsolable realization that they will not return in material form. Instead they will continue to lead an insubstantial existence, however convincing. The phantom limb, rather than a solace as the spiritualists might have deemed the souls of the afterlife, was more often than not a tortuous ordeal.[9] In his catalogue of phantom limb experiences, Mitchell noted that often the last impression of pain continues to goad its sufferer, drawing an analogy with optography, the superstitious belief that the retina of the eye indelibly fixes upon it the last scene which it reflected during life, like a photographic plate: "This fable is realized in the case of lost limbs.

The bent posture of the lost arm is frequently that which it had for a few hours or days before its removal.... The latest and most overpowering sensation is thus for all time engraved upon the brain, so that no future shall ever serve to efface it." ('Phantom Limbs' 568). Such an indexical photographic interpretation of phantom presence is characteristic of spirit photography and the rhetoric of spiritualist phenomena as 'photographic' materializations. With the extrusions of ectoplasmic images, the very body of the medium became at once camera, developer and photograph, so that images of the dead imprinted themselves on diaphanous muslin and issued forth from the medium's orifices.[10] Likewise, the phantom limb is a figuration of photographic memory, capturing the body's image of itself frozen in its last pre-amputation experience.

While spiritualism was engaged in demonstrating the existence of the spirits in the afterlife as a kind of phantasmagorical double of this life clothed in ethereal garments, neurology, especially between the years 1906 and 1935, was in the process of defining another rather different phantasmagorical double, the overlapping yet distinct concepts of 'proprioception', 'body-schema' and 'body-image'. Proprioception, coined by C.S. Sherrington in *The Integrative Action of the Nervous System* (1906), can be described as the body's awareness of itself, its movements and its position. The notion of body-schema was first articulated by the neurologist Henry Head and his co-author G. Holmes in 1911, and then further defined in Head's book *Studies in Neurology* of 1920. It referred to what could be termed the somatosensory 'phantom body', the proprioceptive second body inhabiting and for the most part coinciding with our phenomenological body. For Head, the body-schema is a system of pre-conscious, subpersonal processes that play a dynamic role in governing posture and movement. Body-schema is retentional in that it dynamically organizes sensory-motor feedback in such a way that the final sensation of position is "charged with a relation to something that has happened before" (Head 606). Paul Schilder is the other classic author to have contributed to the definition of body-schema, or, in his words 'body-image', in 1935: "the image of the human body means the picture of our own body which we form in our mind, that is to say the way in which the body appears to ourselves" (Schilder 11). Both Head and Schilder make ample references to phantom limb phenomena as examples of body-schema distortion. In other words, a coherent or semi-coherent body image persists even when the limb has long vanished. In this sense, Mitchell can be seen as the precursor to the body-schema theory,[11] and in *Injuries of Nerves* he gives an accurate and detailed description of the phenomena to which others in later times applied the term 'body-schema':

> We are very competent in health, even with closed eyes, to know where and how far removed the hand may be at any moment, and this knowledge is the result of long-continued and complicated sensory impressions, ocular, muscular, and tactile. Should we lose these by amputation, we cease to have consciousness of the extremity of a limb as set at any fixed distance. (351)

It is as though the newfound awareness of phantom limb phenomena opened up to the description of the idea of the phantom body, the body-schema or body-image. Although not yet employing such terminology, Mitchell interprets phantom limb phenomena, in this case, telescoping limbs, as a glitch of body-schema. This provisional theory, which he apologetically claims is "not altogether satisfactory", is nonetheless supported by the fact that the use of artificial limbs or prostheses serves to correct this unstable telescopic experience, re-embodying the absent limb in material shape and function, and enabling the phantom foot or hand to once again assume its proper place ('Phantom Limbs' 567). Elsewhere, he implies the ghostly pervasiveness of body-schema when describing a telescoping limb which "continues to approach the trunk until it touches the stump, or lies seemingly in its interior, – the shadow within the substance" (*Injuries of Nerves* 351).

One cannot help but see certain parallels (with no claim to precedence or causality) between the notion of body-schema or body phantom – the knowledge of one's own body with the eyes closed, to paraphrase Mitchell – and half a century of experiences of the body in a darkened room during a séance. The theorization of body-image surfaces alongside the cultural practice of heightened perceptive attentiveness in the spiritualist séance, when all those in the room were self-absorbed in the attempt to differentiate their own proprioceptive body-phantom from that of another foreign presence in the room (or that of a fraudulent medium). In the face of disbelief, spiritualism was engaged in an endless rhetoric of substantiation, of tangible proof of invisible forces, despite the fact that the darkened room was anything but a scientific laboratory test site. The mental image of one's own body had to be distinguished from the body which manifested itself as a disembodied hand floating around the room, playing an instrument, writing a note, touching or caressing a witness. Proprioception, as Oliver Sacks would later define it, "implies a sense of what is 'proper' – that by which the body knows itself, and has itself as 'property'" (46). And spiritualism tested precisely these boundaries of self-hood by enabling foreign forces to inhabit and express themselves through this human vessel. The body contained not only the present self, but also the absent phantom, which materialized through the voice or transcribing hand of the medium. Spiritualism and those engaged in antagonizing the movement were faced with the experience of the body as a suffusive permeable receptacle, and attempted to address the blurry susceptibility of both perception of the outer world and proprioception of the inner self. This perceptual vulnerability was undoubtedly one of the main reasons magicians were attracted to tackle the sensory illusions purported by spiritualism. Ectoplasm is perhaps one of the most remarkable instances of these blurred boundaries of inside and out, whereby phantom images or body parts could issue forth from the body of the medium like a photographic second skin or an additional organ. Indeed, in 1925, the Medium Margery, otherwise known as Mina Stinson, mysteriously materialized a hand-shaped ectoplasm of her dead brother Walter, which issued forth from her umbilicus (Cheroux 221). The hand appeared to have grown from her navel, attached to it by an "arm" of sorts, an umbilical parturition that gave birth to a phantom limb. Margery's parasitical phantom hand was a more material manifestation than George Dedlow's phantom legs, which serve not as addenda but as transitory prostheses of re-embodiment. Whereas most spiritualist sittings involved a foreign presence manifesting itself through the body of the medium, Mitchell's story uniquely collapses the phantom body, the séance attendee and the materializing medium into one single corporeal entity: George Dedlow.

Debra Journet suggests that the fictional techniques in 'The Case of George Dedlow' enable Mitchell to explore "a phenomenon that he could not fully articulate within the scientific paradigms of his time" (87). The ambiguities between medical text and fiction are purposefully exploited by Mitchell in the opening lines of his story, where he states that Dedlow's story had been declined by various medical journals, and that although it may not have any scientific value it led him to new metaphysical discoveries (115). The experience of phantom limbs, like the first-hand experience of the converted spiritualist, is one that remains firmly rooted in phenomenological experience, testing the boundaries between the sense of with-in and the sense of with-out the self. Mitchell's story skilfully conflated both the experience of the converted spiritualist and the experience of the most explicit of phantom limbs, phantoms which exceed a neurological explanation to become a supernatural experience. It is possible that Mitchell conceived the term 'phantom limb' concurrently as metaphor and as literal embodiment.

Certainly the ghostly sensations reported by his patients would have stimulated the fiction writer in him, and vice versa.

This chapter has set forth the hypothesis that Mitchell not only dipped into the 'superstitious' beliefs of spiritualism in coining the term 'phantom limb', but that he tested the waters of what was then viewed as an almost inconceivable neurological hallucination through fiction, and later still appropriated some of the sensational rhetoric of spiritualist phenomena to illustrate the extraordinary and seemingly far-fetched truth of phantom limb phenomena in his medical texts. When the author of *The Case of George Dedlow* writes of the 'conviction of its existence', he can be seen to be reaching out to substantiate phantom limbs, and indeed the very invisibility of neural activity as true fictions, so to speak, at the time deducible only through patient descriptions of experienced symptoms and external responses such as reflexes (medical imaging would follow after, enabled by the discovery of the X-ray in 1895). And yet the same turn of phrase, "the conviction of its existence", was equally as pertinent to the fumbling movement of spiritualism, which sought every possible scientific means of authentication to shed light on its supernatural manifestations. 'Conviction' itself implies the defensive need to convince, persuade, convert against the odds, conquer and overcome disbelief. During Dedlow's séance, he is undeniably converted as the medical numbers of his limbs are rapped out by the medium.

> The spelling was pretty rapid, and ran thus as she tapped, in turn, first the letters, and last the numbers she had already set down: "UNITED STATES ARMY MEDICAL MUSEUM, Nos. 3486, 3487." The medium looked up with a puzzled expression. "Good gracious!" said I, "they are MY LEGS – MY LEGS!" What followed, I ask no one to believe except those who, like myself, have communed with the things of another sphere. Suddenly I felt a strange return of my self-consciousness. I was reindividualized [sic], so to speak. A strange wonder filled me, and, to the amazement of every one, I arose, and, staggering a little, walked across the room on limbs invisible to them or me. It was no wonder I staggered, for, as I briefly reflected, my legs had been nine months in the strongest alcohol. At this instant all my new friends crowded around me in astonishment. Presently, however, I felt myself sinking slowly. My legs were going, and in a moment I was resting feebly on my two stumps upon the floor. It was too much. All that was left of me fainted and rolled over senseless. (147–8)

Just as he is 're-individualized' or indeed re-membered by becoming momentarily whole again, his limbs are individualized, 'named' through their respective numerical initials (just as the deceased would manifest themselves at séances through the letters of their names or the dates of their existence). This short story does not yet employ the term 'phantom limb' (they are not named as such until 1871); here instead they briefly materialize as limbs 'invisible to them or me'. All the same, the story can be read as a feat of *naming* possible experience, legitimizing phantom limbs as truly existing limbs inhabiting another realm, that of the mind, and, perchance, that of the spirit-world. Mitchell the medium conjured phantom limbs simultaneously as fiction, as scientific occurrences, and, however much he later denied any allegiance with spiritualism, he undoubtedly also materialized them as proper spirit phantoms.

Notes

1. All subsequent quotations are from the re-published 1900 edition, the two-part book entitled *The Autobiography of a Quack and the Case of George Dedlow*, 113–49.
2. In 'Phantom Limbs', Mitchell describes his fictional account (without admitting authorship), writing "he certainly would have never conceived it possible that his humorous sketch, with its absurd conclusion, would for a moment mislead anyone." (564). He claims the article sets out to rectify those erroneous beliefs. In response to various spiritualist leaders who found in 'Dedlow' confirmation of their methods, he responded sardonically: "Imagine that!" (Mitchell, 'The Medical Department' 1448, qtd. in Goler 175).
3. The 'Phantasmagoria' was a spectacular form of pre-cinematic popular entertainment which involved an atmospheric ghost-show using a magic lantern. The first American phantasmagoria show was in 1803, only four years after its premiere in Paris.
4. From the Old French *fantosme* in the twelfth century, which in turn comes from the Latin *phantasma* ('Phantom', *Oxford English Dictionary*).
5. This was particularly impelled by the neurologist Pierre Paul Broca's 1861 discovery that the functions in the cerebral cortex were anatomically localized, subsequently enabling the diagrammatic mapping of brain functions through the witnessing of reactions to electrical stimulation. That said, a more visual focus was implied in post-mortem autopsy (in animals this was done through the studying the effects of surgical lesions).
6. This is not the context in which to expand on parallels with hysteria, but I will mention briefly that spiritualism was often pathologized as a mental illness by hostile male authorities of the medical profession, and the specifically feminine demonstrations of mediumship were interpreted as manifestations of hysteria (Braude 157). It is also notable that the twitching spasms and automatic gymnastics of the stump caused the male amputee

to be perceived as an effeminate hysterical body (cf. O'Connor, 'Fractions of Men' 744).

7. Earnest sees in Mitchell the beginnings of Freud's psychoanalytic method (227).

8. Mitchell was well-known for his invention of the so-called 'rest cure', initially developed to treat soldiers with battle fatigue, but frequently applied to women with nervous symptoms such as neurasthenia and hysteria. It was enthusiastically taken up in both America and Europe, and involved enforced rest, seclusion, controlled diet in the form of excessive feeding, massage and electricity. It was administered to women such as Charlotte Perkins Gilman, who condemned both Mitchell and his treatment in her famous 1892 autobiographical story *The Yellow Wall Paper*. Mitchell allegedly altered his rest cure after reading Gilman's account (cf. Thrailkill, 'Doctoring "The Yellow Wall Paper".')

9. A typical (and mostly unsuccessful) treatment for the relief of phantom limb pain was to re-amputate, as it were, the phantom, by repeatedly shortening the stump further up *(Injuries of Nerves* 286).

10. The medium Eva C. was the most famous to photographically materialize such images whilst also being documented in photographs. The resulting images taken by Schrenck-Notzing were published in 1913 in Germany under the title *Phenomena of Materialisation*, and translated into English in 1920 (Warner 232).

11. So he is deemed by Tiemersma (110).

Works cited

Braude, Ann. *Radical Spirits: Spiritualism and Women's Rights in Nineteenth-Century America*. Bloomington: Indiana UP, 2001.

Brown, Edward M. 'Neurology and Spiritualism in the 1870s.' *Bulletin of the History of Medicine* 57 (1983): 563–78.

Castle, Terry. 'Phantasmagoria and the Metaphorics of Modern Reverie.' *The Female Thermometer: Eighteenth-Century Culture and the Invention of the Uncanny*. New York: Oxford UP, 1995. 140–67.

Cheroux, Clement, Andrea Fischer et al. *The Perfect Medium: Photography and the Occult*. New Haven: Yale UP, 2004.

Cottom, Daniel. *Abyss of Reason: Cultural Movements, Revelations, and Betrayals*. New York: Oxford UP, 1991.

Earnest, Ernest. *S. Weir Mitchell, Novelist and Physician*. Philadelphia: Pennsylvania UP, 1950.

Freud, Sigmund. *On Aphasia: A Critical Study*. Trans. E. Stengel. London: Imago, 1953.

Goler, Robert I. 'Loss and the Persistence of Memory: "The Case of George Dedlow" and Disabled Civil War Veterans.' *Literature and Medicine* 23.1 (2004): 160–83.

Head, Henry. *Studies in Neurology*. London: Henry Frowde, Hodder & Stoughton., 1920.

Herschbach, Lisa. ' "True Clinical Fictions:" Medical and Literary Narratives from the Civil War Hospital.' *Culture, Medicine, and Psychiatry* 19.2 (1995): 183–205.

Journet, Deborah. 'Phantom Limbs and "Body-Ego": S. Weir Mitchell's "George Dedlow." *Mosaic* 23.1 (1990): 87–99.

Keen, William M., Silas Weir Mitchell, and George R. Morehouse. 'On Malingering, Especially with Regard to Simulation of Diseases of the Nervous System.' *Philadelphia Medical Journal* (October 1864): 364–94.

Kittler, Freidrich A. *Gramophone, Film, Typewriter.* Trans. G. Winthrup-Young and M. Wutz. Stanford: Stanford UP, 1999.

Long, Lisa A. *Rehabilitating Bodies: Health, History and the American Civil War.* Philadelphia: Pennsylvania UP, 2004.

Louis, Elan D., Stacy Horn, and Lisa Anne Roth. 'The neurologic content of S. Weir Mitchell's fiction.' *Neurology* 66 (2006): 403–7.

Mitchell, Silas Weir. *The Autobiography of a Quack and the Case of George Dedlow.* New York: Century, 1900.

——. *Injuries of Nerves and their Consequences.* 1872. New York: Dover, 1965.

——. 'The Medical Department in the Civil War.' *JAMA* 52.19 (1914): 1445–50.

——. 'Phantom limbs.' *Lippincott's Magazine of Popular Literature and Science* 8 (1871): 563–9.

——. 'The Treatment by Rest, Seclusion, Etc. in Relation to Psychotherapy.' *The Journal of the American Medical Association* 50 (1908): 2033–7.

Mitchell, Silas Weir, George R. Morehouse, and William M. Keen. *Gunshot Wounds and Other Injuries of Nerves.* Philadelphia: J. B. Lippincott, 1864.

O'Connor, Erin. ' "Fractions of Men:" Engendering Amputation in Victorian Culture.' *Comparative Studies in Society and History* 39.4 (1997): 742–77.

——. *Raw Material: Producing Pathology in Victorian Culture.* Durham: Duke UP, 2000.

Paré, Ambroise. *Oeuvres completes d'Amboise Paré.* Vol II. Ed. J.F. Malgaigne. Paris: Baillere, 1840–1.

'Phantom.' *Oxford English Dictionary.* 2nd ed. Oxford: Clarendon, 1989.

Price, Douglas B. and Neil J. Twombly. *The Phantom Limb Phenomenon: A Medical, Folkloric, and Historical Study. Texts and Translations of 10th to 20th Century Accounts of the Miraculous Restoration of Lost Body Parts.* Washington DC: Georgetown UP, 1978.

Sacks, Oliver. *A Leg to Stand On.* London: Picador, 1991.

Schilder, Paul. *The Image and Appearance of the Human Body.* London: Kegan Paul, Trench, Trubner, 1935.

Sconce, Jeffrey. *Haunted Media: Electronic Presence from Telegraphy to Television.* Durham: Duke UP, 2000.

Thrailkill, Jane F. 'Doctoring "The Yellow Wall Paper".' *ELH* 69.2 (2002): 525–66.

Thurschwell, Pamela. *Literature, Technology and Magical Thinking, 1880–1920.* Cambridge: Cambridge UP, 2001.

Tiemersma, Douwe. *Body Schema and Body Image: An Interdisciplinary and Philosophical Study.* Amsterdam/Lisse: Swets & Zeitlinger, 1989.

Warner, Marina. *Phantasmagoria.* Oxford: Oxford UP, 2006.

Weinstein, Sheri. 'Technologies of Vision: Spiritualism and Science in Nineteenth-Century America.' *Spectral America: Phantoms and the National Imagination.* Ed. J.A. Weinstock. Wisconsin: Wisconsin UP, 2004. 124–40.

6
Modernism and the Two Paranoias: The Neurology of Persecution

George Rousseau

> Sorrow begins in preoccupations with the self
>
> Ovid, *Tristia*, III.IV.5

I

A rather wild, recent book by British artist–photographer Ralph Steadman illustrates one hundred 'famous paranoids' from Socrates to Joan Collins. Each figure possesses a grotesquely extended nose, twitching and disfiguring the subject's face through distortion. Steadman naughtily suggests that these celebrities are 'paranoids' because society cannot deal with abnormally large noses or freakish ones incapable of containment. An odd intimation when he means something else ... but hoping to keep his cards close to his chest regarding prurience and fame he wittily passes them off as 'paranoids', his cunning label for their common status as front-row celebrities.

Steadman's 'twitchers' include Jesus and the Virgin Mother, the British Royals (Queen Elizabeth II and the late Diana Princess of Wales), every type of Hollywood glitterati (Charlie Chaplin to Marilyn Monroe) and artists (from Bach and Mozart to Margot Fonteyn and Nureyev). If there is an intellectual point to be made here it is that stereotypes of paranoia are fundamentally visual rather than verbal: stereotypes are pictorially mentalized. And Steadman locates their recent incarnations – "paranoid moderns" as he calls them – at the end of his book, in three pages of pictures *sans* verbal text entitled 'Paranoids'. These last pages contain ten illustrations displaying 'God with ten faces', especially God's own extended nose, fluttering and twitching. The book closes with equally grotesque images of a nasal Devil, and finally with the foulest nasal grotesques of them all, Adam and Eve.

Steadman's subtext is the *nervous nose*: fluttering so profusely it epit-
omizes the pith of celebrity. He knows that within human anatomy
the erotogenous zone in the genitals – the penis and vaginal cavity –
contains the body's greatest concentration of nerves outside the brain:
millions of overlapping nerve cells affording human beings the 'big
bang' that made sex the *sine qua non* of experience long before Sigmund
Freud memorialized it. Pleasure, unadulterated pleasure, connects Steadman's paranoids: the cupola of celebrity and sex. What else is celebrity,
he asks, especially when fanatically chased as it is now, than the fantasy
that when we attain it every pleasure – especially sexual pleasure – will
beckon at our feet?

In a far less satiric mood American literary critic John Farrell, in *Paranoia and Modernity*, has observed the degree to which paranoia has
infiltrated modern sensibility.[1] "The dominant figures of modern culture exhibit a strange susceptibility to delusions of grandeur and fears
of persecution upon imaginary grounds – in other words to paranoia",
he suggests. These words open his survey of paranoia in modernity. The
exploration is intuitive: "the inquiry that led to the writing of this book
began as an attempt to understand the paranoid slant of modernity"
(*Paranoia* 1). Yet no sooner has Farrell entered paranoia's labyrinth than
he discovers its inhabitants lurking: Henry James, Nietzsche, Strind-
berg, Ibsen, Maupassant, E.T.A. Hoffmann, Schopenhauer, Stendhal,
Dostoevsky and James Joyce; almost all American novelists, including
Melville, Burroughs, Mailer and Pynchon – modernism's extended family. The list is long. Read Farrell's book and you must believe that every
artistic figure since the late nineteenth century was paranoid: paranoia,
the illness of the modern.

Farrell also notices that modernity's paranoia "clearly sprang from the
seventeenth century and the transformation of medieval into modern
intellectual culture" (*Paranoia* 2). The observation rests on the sociological theory of R.H. Tawney and Ernst Troeltsch about the Protestant
work ethic and the rise of capitalism; an heroic attempt by Farrell but
too sweeping and lacking in historical depth. Farrell never defines his
key concept: paranoia remains a loose catchall – a wide umbrella – for
every form of otherness: madness, delusions and persecution, eventually embracing everyone. More nuanced is David Trotter, who seeks to
establish what paranoia actually meant for modernism. Like the wolf,
Trotter works by stealth: one step forward, two steps back. No sooner
has he offered up paranoia as modernism's insignia than he strains
hard to qualify it: "By no means all the writers I would associate with
paranoid Modernism were English, or male, or novelists." Perhaps they

were not but Trotter refines paranoia's national, ideological and gender bases even further: "paranoid Modernism was markedly – though by no means without exception – English, male, and novelistic." The notion is controversial and gives even the least erudite students cause to pause (Dostoevsky, Kafka, Robert Musil, English?). Yet Trotter challenges his own claim again: "But it·[paranoid Modernism] was only ever one Modernism among several" (12).[2] Trotter's impressive argument succeeds by the controls placed on the logic of paranoia itself, especially its genealogy and cultural history.

"Paranoia, too, has a history", Trotter claims, whose broad genealogy he wants to reclaim: "it has always been a diagnosis made from contexts." Farrell was cavalier about them but Trotter cautions students further that "to denaturalize it [paranoia] is to begin to understand its meanings, then and now." So its profile is not transhistorically glimpsed and its genealogy is ancient, transformative and protean, as Trotter claims, "as yet almost entirely unwritten in the wider social and ideological sphere". Yet even Trotter succumbs to the temptation of generality despite his valiant "brief history of paranoia", by neglecting its cavernous rupture in the middle of the nineteenth century (17). What results is a somewhat potted paranoia which, however modulated, minimizes the slow transformative process that eventually changed it from broad delusional madness to specific persecutory mania. The process has deflected historians of medicine and psychiatry for a generation and continues to vex them now.[3] Its late nineteenth-century transition from delusional mania to persecution is the subject of this chapter, especially the neurological components of that shift, and I want to commence discussion of the shift, as well as the curious growth of the new persecutory paranoia and the rise of modernity.

II

The old classical paranoia, as even the simplest lexical trajectory of the Greek word demonstrates, denoted *all* types of madness: the most fleeting delusion as well as ephemeral obsessive thoughts. Yet paranoia was never the whole of madness, merely its delusional sector; the cognitive dimension infiltrating awareness rather than influencing behaviour, and as such was especially prone to religious extremity. It was the version of paranoia, for example, that sent the critics of Puritanism and other ranting sects of the early modern period to scrub them as 'mad'. Even Jonathan Swift excoriated them in his *Tale of a Tub* (1704) for their delusions and hallucinations.[4] The medicalization of madness in the mid-eighteenth century challenged the old paranoia, its 'mad doctors' (as the new theorists of insanity were then called), generating diverse theories to pinpoint its 'mechanical operations'. Just as Swift had ridiculed the deluded mad in his *Tale*, often drubbing Newton's methods as well as casting doubt on mathematical Newtonianism, the early Enlightenment physicians established models to demonstrate that paranoia was a derangement of the body's equilibrium in fluids and solids. By the end of the eighteenth century the old designation appeared too flaccid and imprecise to satisfy anyone – something was destined to refine it.[5]

The rise of nervous science challenged the old paranoia and forced it to ask hard questions. Ever since the 1680s, when the followers of the great neurologist Thomas Willis (including Samuel Pordage, who translated Willis's work from Latin to English in 1680) began to achieve success in disseminating Willis's notion that the brain was the centre of the body's operations – for intellection, mental sanity and insanity – thought and action began to be explained through neural pathways. The Scottish Enlightenment Newtonians perfected its demonstration: William Cullen, the most influential Scottish medical doctor of the Enlightenment, taught that the nervous body was a microcosm of nervous society; on this body could be scripted the culture's vitality: the healthier the human body the more permeated with nerves. Likewise and by analogy the body of the civic polity. A tautly nervous human body indicated mental health, its opposite – flaccid, languishing nerves – depression and even lunacy in extreme cases. Follow this logic to its conclusion and it becomes evident that insanity develops from anatomical–physiological states in which deranged, weak nerves acting on a disordered brain alter the individual's rational powers.[6]

Circa 1800, as the German *natürphilosophien* constructed their theories of Romanticism and Coleridge and Wordsworth penned their revolutionary 'Preface' to the *Lyrical Ballads* instaurating a function for literature, signs that the old paranoia was exhausted were arising. For example, William Godwin, a radical figure and author of the last totalizing Enlightenment discourse about political justice (his 1795 *Political Justice*) can legitimately be called 'paranoid' in the old classical sense: biographically delusional, his cognitive faculty distorted and his physical body possessed by bouts of epilepsy. *The Adventures of Caleb Williams* (1794), his fictional exploration of a nervously possessed murderer, explains Falkland's paranoia by demonstrating how a lifetime of repetitive nervous experiences leads up to the murderous epileptic fit that induces him to kill, introducing the persecutory element as coeval with the delusional. The autobiographical element is patent: during these years, when Godwin was writing *Caleb Williams*, he himself experienced a gradual rupture from his old rational mindset about the possibility of a coherent social order to the nervously sceptical epistemology that would later characterize early modernist writing. Concurrently, Thomas Trotter, the leading naval physician in Nelson's Fleet, explained how the body's nerves inscribe social patterns over long stretches of time that could be decoded by the psychiatrist's keen scent; the trail of an ailing social polity producing deranged nerves in the bodies of its citizens.[7] The nerves do this by repeatedly hard-wiring experience into the nervous system. Trotter's sense of the physiology of the process was minimal, but his notion of hard-wiring through experiential repetition has been borne out by the best neuroscience. Also concurrently, leading French psychiatrist Jean Esquirol spent a long career in Paris treating patients who thought they were Napoleon or Robespierre: persecuted as well as strangely triumphant in their inner psychic life (II: 98). And James Cowles Prichard, the English medical anthropologist who coined the term 'moral alienation' to describe what he understood as yet another new form of mental insanity, was also dissatisfied with the old models of paranoia. Writing in 1822, he claimed that the 'paranoid' were obsessed but had minds still perfectly intact. Prichard's theories stirred fierce professional controversy among psychiatrists and legal thinkers who believed that the deranged had compromised their intellects.

There was much debate about these views in the early nineteenth century but eventually even the influential Esquirol agreed with Prichard. However disparate the models of these contemporaneous theorists, they were nevertheless unified in the belief that derangement was not owing entirely to physiological defect – possibly acquired through heredity – but generated by social disorder. Its imprint on the nervous system of the sufferer was the *physical* sign of its presence. Such a new homology linking mental illness to the fabric of social disarray required a new model of insanity: if the old, depleted paranoia were to survive something was needed to regenerate it.

Yet much European medical education after 1800, especially in the great teaching hospitals in Paris, Vienna and Berlin, challenged the old paranoia by demonstrating that *all* delusional thinking, not merely certain forms, originated in derangements of the nervous system. Years of experimentation, played out as theatre in its lecture rooms, showed the effects of these altered nervous states. Here, in the lecture halls of the Salpêtrière and Bicetre presided over by Pinel and Esquirol and eventually Charcot, the demonstrations were made – the same lecture theatres where the German and Austrian psychiatrists of the next generation were being educated: doctors Georges Falret, Carl Westphal and Kraft-Ebbing who would change the face of the old paranoia forever.[8]

But why should the old paranoia – a network of delusions not yet labelled 'megalomania'[9] – have begun to give way to a new persecutory insanity? Whatever the causes, the transition was gradual and extended over a century. Its decade-by-decade patterns remain to be charted by historians of psychiatry intimately familiar with the subtlest nuances differentiating them. Certainly the old religious mania, sometimes called melancholia, could also present itself with persecutory aspects. More noteworthy is that the early nineteenth-century psychiatrists began to recognize the part played by social disorder – especially widespread political unrest and upheaval – in the formation of insane persons. After the cataclysmic events of the 1790s, Esquirol was not surprised that his megalomaniacal patients imagined themselves to be Napoleon; after the revolution of 1848 and its dire fallout throughout Europe, it was expected that the persecuted insane would pretend they were Tallyrands and Metternichs; and they could do so often without the slightest trace of diminished insight or intellectual capability. Slowly, persecutory madness made strides into religious and delusional insanity concurrently with a European social order that was less stable than it had been for decades.[10] Hence set the dials to 1850, or 1860, when Baudelaire forged a new aesthetic eventually called 'modernism' and Poe and Melville toiled on their own versions independently across the ocean, and a new type of paranoia is also creeping in. It affirms that paranoia presents itself as *persecutory insanity* rather than in merely delusional or megalomaniac forms, but its detection is slow: psychiatric thought did not suddenly discard decades of belief, as does the lizard when shedding half its body under attack.

The growing explanation was that the deranged nerves of the persecuted were symptoms of a disintegrating social fabric, one witnessing brutal uprisings among nations (Europe's fragile power relations after 1815 and 1848), fragmenting religions (the crisis in Catholicism as the Hapsburg empire tottered) and catapulting the social classes (Marx and communist manifestos). Still, the nineteenth-century psychiatrists were not social scientists; at best their neurology and psychiatry reflected trends observed in their patients: the persecution they observed in their waiting rooms. But why this chimera and why persecution rather than infatuation, stealth or other cunning? Why the pervasiveness of persecution among the lunatics and malingerers they examined? When looking backwards from the post-Second World War twentieth century it may

appear preordained that persecution – the type that led to its medicalization and the diagnosis of schizophrenia that soon exploded after the
Great War – would become the modern's dominant emotion. But set
the dials to 1848 or 1871, or any other perilous political moment before
the *fin-de-siècle*, and the future of paranoia is more difficult to gauge.

Moreover, no scholar has ever made an airtight case for the con-
currence of European Romanticism and the rise of psychiatry (as a
professional practice as well as a way of knowing human beings), but
students of these matters could do worse than detect the origins of
the new paranoia among the Romantics – in their literature as well
as their psychiatry. Insightful Romanticists have spotted some of the
anticipations but more discussion and evidence is needed.[11] Here we
return to Farrell and Freud – and we might add Jürgen Habermas in *The
Philosophical Discourse of Modernity* about the failure of the great Enlightenment ideologies in the early nineteenth century – as well as the view
that "paranoia was a psychological substitute for religion" (Farrell, *Paranoia* 3). The argument about this substitution of a surrogate religion, as
many historians have observed, is too facile, but Farrell's tack is more
promising. He shrewdly observes that Freud himself displayed the traits
he labelled as 'paranoid' in others. 'Freud as Leonardo' and 'Freud as
Michelangelo' have struck others than Farrell.[12] Still, Farrell's passage is
worth citing in full:

> For Freud … we all participate in paranoid thinking: our way of mak-
> ing sense of things is to him a manifestation of illness akin to
> paranoia: 'The moment a man asks about the meaning and value of
> life, he is sick, since objectively neither has any existence.' This state
> of affairs causes Freud no discomfort. He appears to enjoy likening
> himself, and the rest of humanity, to the paranoid.… Freud's para-
> noid has the character of an overly credulous intellectual, a religious
> zealot who can neither distinguish wish from reality nor preserve
> the proper level of scepticism toward his own thought. Freud knows
> himself to be smitten with the same curse as the paranoid, to be an
> irrational being of the same kind, only he, as a scientist, has kept his
> credulous tendencies moderately in check. What distinguishes the
> scientist from his paranoid patient is not so much a superior under-
> standing of the world, or even of other people's motives, as much
> as a superior awareness of his own irrationally self-aggrandizing
> nature. It is the difference between naïve and sentimental paranoia.
> (*Paranoia* 3)

As if this convergence were insufficient Farrell also seizes upon Freud's megalomania as the sign of his own paranoia (and he might have buttressed his argument by citing Freud's increasingly strained letters with Fliess):

> Freud shares with the paranoid character an unabashedly heroic self-image tending, as he admits, toward megalomania; he sees the social world as being fundamentally hostile and as threatening specifically toward him; he excels at finding hidden, malignant significance in the behaviour of others; above all, he is systematically suspicious. (*Paranoia* 3)

III

Yet if Freud is viewed as too subjective in the transition from the old to the new paranoia, to whom are we to turn with confidence? One alternative minefield is literature and philosophy: those large hulks of early modernist thinking that commented on the nature of reflection without worrying about Freud's biographical 'status anxiety' (the phrase is Alain De Botton's). Surely, however, it is myopic to discard Freud altogether, especially the Freudian part that imported neurology into his thinking about paranoia. He chose neurology as his specialty within medicine, spent years aiming to reduce the paranoid human personality to its neurophysiological bases and to producing a "Project for a Scientific Psychology" that would establish "a psychology that shall be a natural science" (295). Later he launched his psychoanalysis after educating himself in Charcot's hysteria-laboratory in Paris by combining neurology and physiology to forge the new psychology.[13] Of course we want to establish a reliable and historically nuanced "theory of paranoia", as David Trotter has mandated (17); one less reliant on Freud's corpus of writings about paranoia, as in the sadly wrongheaded essay about Dostoevsky's neurology.[14] Such a theory will extend beyond Freud to the new paranoia – a broadly based *verrücktheit* or persecution – appearing with frequency in psychiatric literature by the mid-nineteenth century long before Freud pronounced on paranoia.[15]

But the discursive implications of the new paranoia must also figure in, and they resist literal charting because segments of their existence had been implicit in the old paranoia. As Peter Logan, an astute student of 'nervous narrative', has noticed, nervous illness itself led to narrative voice: "without the disease there would be no narrative, not even one with the social utility of warning against the social conditions that created it in the first place." Logan therefore privileges disease over narrative. He means that malady is antecedent to discursivity, in our case that persecution itself gave rise to paranoia's voices, whether scripted or heard inside the head. Logan's further virtue is that despite his separation of the two strands – disease and narrative – he thoroughly resists ranking them hierarchically (3). Two cruxes therefore present themselves: the sources of the new paranoia and its discursive elements. Logan presents a third in the ambivalence in which the speaker strove to gain authority in his speaking voice despite the presence of his ailment, while recognizing that the nervous malady has also empowered its resulting narrative. Logan puts it this way: "paradoxically, the nervous narrative promotes, in its formal structure, the same disorder it cautions against by transforming the narrator's debility into a narrative premise. This problem only intensifies as the narrative gets increasingly convincing" (79).

Logan writes primarily about hysteria rather than paranoia (let alone the *new* paranoia), but his form of nervous ailment poses no conceptual hurdle so much as identifying a persistent interpretative tension at work in *all* nervous narratives: hysterical, hypochondriacal and paranoid. For the same wide-ranging and unrestricted signs applied to the forever-protean hysteria also apply to paranoia. Both conditions were configured as inherently unstable, uncontainable, flexible *ne plus ultra*; stealthily adapting to different settings of time and place; incapable of sustaining the localized contexts that could permit psychiatrists to generalize about their genesis.[16] Consequently both were deemed inscrutable throughout the long nineteenth century. And both proliferated beyond expectation in hospital admissions, with the exception that the new persecutory paranoia – *verrücktheit* – began to suggest a handle on which to lodge the belief among model builders, by virtue of its concrete anchor in persecution, that the new paranoia would eventually be explained by progress in 'nervous science'.[17] Here, in the 1870s and 1880s, was the expectation of a breakthrough in modernist neurology: the baggage the young Freud imported into his later thinking, as we shall now see.

However, my allusion to 'breakthrough' may appear controversial if construed to indicate that *all* neurology is 'modernist'. It emphatically is not, and in this chapter I have sought to demonstrate that while neurology (i.e. writing about the nerves) at large was an ongoing province of inquiry since the Greeks, magisterially reinvigorated in Europe by physiologist Thomas Willis in the late seventeenth century,[18] some branches of post-1830 neurology participated in the same debates as the early proponents of an emerging modernism. For example, the idea that 'the soul is neurological' was old by the mid-nineteenth century, but the neurology of fear and shock rather newer. The interface of neurology and modernity has been central to my concerns here. If there had been space we could have probed which tenets and proponents of nineteenth-century neurology can legitimately be called 'modernist' in the senses I have been using the term. But the demonstration can also seem gratuitous and even pedantic, for the point about the new persecution is that once it was culturally manifested its neurophysiology required attention. Historically the case was not the other way around: a neurology of persecution arose and culture at large somehow found a use for it. Put otherwise, segments of the burgeoning discourse of mid-nineteenth-century neurology – especially components of its physiology – can fairly be called 'modernist' because they dovetailed with culture in general. I write "segments of the burgeoning discourse" because my approach focuses on demonstration of how the new paranoia diagnosis conformed to the general persecutory components of late nineteenth-century society.

To step further back historically once again, if reflection and the reflexive habit of mind indicated a tendency towards paranoia – especially to persecutory anxiety as David Trotter argues – it was also necessary to explain the enigma of its origins. How did it arise in people? When Enlightenment theorist Boissier de Sauvages revived the old paranoia in the late eighteenth century as important for his classification of mental diseases accompanied by delirium and fever, he was silent about 'reflective minds' as prone. Nor was any anatomical expertise needed by the mid-nineteenth century to demonstrate that sensation was sufficient to form experience (the doctrine of innate ideas had long ago been smashed and even its vestiges among religious thinkers quashed). Sans sensation and its impressions made on the nerves, no experience could be remembered, let alone constructed. By the 1880s magazines and newspapers were claiming that the "quality of modern life" depended upon "the daily sensations of ordinary life".[19] But sensation also lodged in the heart of many late Enlightenment – especially in Condillac's influential works – systems charting experience.[20] Why then was the old paranoia, which was chronologically commensurate with theories of neurological sensation circa 1800, not sufficiently *reflexive*? Why was the new persecutory paranoia of the nineteenth century needed to establish reflexivity as the culprit of mania?

The riddle cannot be unravelled here – an entire book is needed. What can be affirmed is that the body was not yet (in c.1800) viewed as *sufficiently* inscribing experience through sensation onto the nerves. It may have been thought to be so doing in certain anatomical and medical treatises, but their learning had not yet seeped into popular imagination.[21] Not even versions of 'Gothic' sensibility, which thrived on exalted states of the nerves aroused by awesome imagery, did so, despite external signs that it was capable.[22] When viewed retrospectively the future of neurology circa1830 indicates that the nervous system was still young, so to speak.[23] It had not yet matured as it would in the new neurology of the second half of the nineteenth century. This solution to the riddle points unabashedly to the neurology commensurate with modernism.

Stated syllogistically, bodies insufficiently nervous are (1) unable to reflect social disorganization accurately and (2) also incapable of displaying (Logan's) tension over anxiety and empowerment – a double-headed impediment. Once configured as thoroughly nervous bodies, as they will be in the mid-nineteenth century, they can take off in both ways and the evidence demonstrates that they did. Survey the late nineteenth-century naturalistic novel, for example, and both so-called impediments evaporate. Zola, for example, thoroughly exploited the nerves to develop the psychological naturalism of his novels. Virtu-ally each of the paranoid characters in *Thérèse Raquin* (1867) comments on the fractured states of his or her nerves, protagonists Laurent and Thérèse many times. We expect the sickly Camille, her honest and naively sentimental husband, to have weak nerves but are surprised when the others – in various degrees paranoid about Camille's death – comment so explicitly. Laurent and Thérèse, the adulterous killers, have good reason to be acutely paranoid. They live in daily dread of the discovery of their drowning of the hypochondriacal Camille. Yet Zola cleverly calibrates their nervous states as a secure guidepost to their post-murderous moods: at first recklessly doting on each other, then nervously devouring each other, and finally committing double suicide by drinking poison. Their morbid nervous states can only have func-tioned so well in his narrative if Zola's audiences had been previously prepared to recognize them as signs of heightened paranoia.

IV

Another paramount reason exists to retain Freud. We recall his insis-tence that "I shall not consider any theory of paranoia trustworthy unless it also covers the hypochondriacal symptoms by which that dis-order is almost invariably accompanied."[24] Whether hypochondriasis was the major symptom depended, for Freud, upon each clinical case. But in this capacity he represented the end of a wave, not its beginning. Much earlier, the late Enlightenment doctors had issued their own dicta about the centrality of hypochondria to nervous cases. Dr. John Hill, then a prolific if now an obscure author about 'new diseases', writing in *Hypochondriasis* (1766), one of the earliest treatises explicitly focused on hypochondriasis, chose it above all other symptoms because he had observed that the majority of nervous disorders present with it. And Swiss Protestant physician Samuel Tissot dwelled upon it because he noticed that literary people selected it as their chief complaint when consulting him. Godwin, already discussed, appeared to his contemporaries the perfect prototype of the Tissotian gaze, as did the fictional Falkland, the paranoid murderer in *Caleb Williams*. Godwin's enemies often claimed that his entire career had amounted to the demonstration of a sustained case of persecution, as I suggested earlier, and there is little doubt that his biographical paranoia dovetailed with the modernist tendencies of the mid-nineteenth century.

Continue in this vein, decade by decade, and it can be seen how prolifically the Romantics added to the gospel about self-aggrandizement even if it was not yet labelled megalomania: the several volumes of *Memoirs of an Invalid* (1819–20) by Henry Matthews, an inferior author who imagined that his malady mandated him to write; Shelley's later poetry; Robert Schumann the hypochondriacal suicide; Nietzsche's forms of self-aggrandizement and, paradoxically, his comments on the social utility of illness; Henri Amiel's pathetic belief that he had been ordained by the gods to monitor his nervous states each and every hour and, as a result, produced the longest diary in the French language; Dostoevsky's life of 'possession' and his novels reflecting the frenzied states of Nicolai Stavrogin whose paranoia has been shaped by the political order; composer Richard Wagner's commission of a whole series of dastardly acts while retaining his exquisitely logical capability in the composition of *Parsifal*, itself the myth of an ailing 'Society of Brothers' in search of the Grail; down through the decades to Flaubert's hidden sickness, Ezra Pound's mental malady and Robert Musil's forms of alienation.[25] Glancing backwards, if much social coherence was gone by 1848, never again to be restored, then the shattered nerves of its artists – like these just mentioned and many others – were living proof.

Is there any wonder then that the intuitive Farrell could decide to spend a working career on 'paranoia and modernism' despite a thin sense of paranoia's recently nuanced medical past? Small surprise that he was struck at how Henry James imagined himself as Napoleon and

Nietzsche as Augustus Caesar. Farrell's opening words merit 'citation' again, even if cited here in truncated form: "The dominant figures of modern culture exhibit a strange susceptibility to delusions of grandeur and fears of persecution upon imaginary grounds – in other words to paranoia" (*Paranoia* 1). They do indeed: megalomanias as well as the persecuted. Other critics before Farrell noticed to what degree paranoia was the condition of modernism *par excellence* even if they rarely refined it, as Freud did, into branches delusional, megalomaniac and hypochondriac. If Daniel Schreber had not been hypochondriacal, Freud would not have elevated him to one of history's most celebrated (in Steadman's sense) paranoiacs. These strange nineteenth-century convergences of paranoia and the rise of modernism merit sustained comment, not because paranoia's symptomology is clinically unique, or because one causes the other (paranoia and modernism), but rather for the ways the illness serves the anxieties inherent in modernism.

In medical theory of the day hypochondria was indeed the indispensable *sine qua non* symptom of paranoia. But the more one scrutinizes Freud's *oeuvre* the clearer it becomes that hypochondriasis is merely the sign of something *else*: of the fundamentally nervous aberration in the body of the persecuted. Did Freud then see himself as the tormented Dostoevsky? Some critics have thought so and elevated the similitude to a view of the essential Freud. Even Farrell, having detached himself from a reading of the Dostoevsky essay, comments in the passage quoted above that Freud "appears to enjoy likening himself... to the paranoid". Freud may indeed have "revelled" in the irony of likening himself to Dostoevsky, and the sustained late twentieth-century attack on him was conceived and executed, in part, as inquisition of the Promethean megalomaniac thinking that he could pierce to the darkest secrets of humankind. But the irony is double. If Freud could have returned, like Lazarus, from the dead and witnessed the savage attacks on himself, he would have redoubled his view that paranoia – the new paranoia attendant with hypochondriasis and megalomania – is indeed the condition of the modern. He himself had been living proof.

By the time Freud died in September 1939 persecutory paranoia was starting to play itself out, or at least dissipating itself into bourgeois sensibility, despite the Allies' declaration of war against Germany. But not without modernism's attendant paranoia slowly heading south. Set the dates earlier, approximately to the clouds of war gathering in 1914 in America or the crash of 1929, and paranoia – the unjustified excessive sense of fear exhibited in *verrücktheit* – is doubtless the major psychological sign of the modern. America's then greatest living historian, Arthur Schlesinger, would ordain it had been, as would critic Leo Strauss in discursive contexts.[26] Yet in Europe its moment ebbed. As Nazi insanity increased after Freud's death and the secret about concentration camps could no longer be contained beyond Germany's borders, society was racked by something far more horrendous than an "unreasonable fear of the actions of others", as the dictionary definition of paranoia was then commonly found.

Hitler's excesses topped the realm of the 'unjustifiable'. By the 1940s 'unreasonable fear' of the moderns was extended to whole political parties, governments and 'Big Brother' coming to devour you – as Orwell wrote in 1948 – in *Nineteen Eighty-Four*. Vladimir Nabokov epitomized the shift at the end of the War when observing that paranoia's infiltration into the whole fabric of society had been so thorough that "there is nothing more banal and more bourgeois than paranoia" (IV: 96.) It seemed so to him. Despite the decline and absorption into mainstream culture, paranoia's capital in nerves never really diminished. The new schizophrenia, about which the late Gilles Deleuze and Félix Guattari wrote so eloquently in *Anti-Oedipus: Capitalism and Schizophrenia* (1984), was equally 'nervous', but one seems to have replaced the other in the genealogy of persecution, and by the 1960s, when auguries of postmodernism could be heard from the likes of Baudrillard and Lyotard, the talk was of schizoid cultures splintered by their anxiety over the genuine and the fake, the authentic and the simulacrum. Yet the new schizophrenic prototype of postmodernism was as physically 'nervous' as his modernist forebears had been. Persecution's forms rose and fell, but its neurology remained a constant. Perhaps the naughty Steadman scored a point after all when illustrating it in those twitching noses from Socrates to Joan Collins.

Notes

1. Earlier Farrell had examined Freud's versions of paranoia in *Freud's Paranoid Quest*.
2. This work opens with a 60-page 'Brief History of Paranoia'.
3. See Lewis, Berrios and Berrios and Porter. The most useful assessment of the history of the concept is found in Dowbiggin.
4. See Smith, and, for its modern vestiges, Carroll.
5. See Lewis for the medical background.
6. For more detail about the development of the homology see Rousseau, and Bynum.
7. Trotter's chief social ailment was drink, producing widespread alcoholism corrupting individual bodies; the habit arose from larger dissatisfactions with modern life.

8. Westphal had written one of the earliest treatises on the new persecutory *verrücktheit* in 1878. But Emil Kraepelin put paranoia on the map and twice summed up its theory during his late nineteenth-century youth (see Kraepelin and Diefendorf, and Kraepelin and Robertson).

9. In English 'megalomania' was used with some regularity by the 1880s, often as a synonym for the much older 'monomania': any obsession with compulsive, repetitive thoughts, as German phrenologist J.G. Spurzheim observed as early as 1815. Neither Romantic monomania, nor the much later megalomania, specifically denoted the persecutory aspect. These terms overlapped in the psychiatric theory of the nineteenth century and rarely are found in isolation from each other.

10. The two were commensurate rather than causal and it would be premature, if not false, to claim that one (political instability) caused the other (persecutory insanity). Mark Micale also yokes political history and the rise of psychiatry (see Micale and Lerner). Others point to shifts in gender stability as contributory (see Gordon).

11. Thomas Pfau, for example, has also commented perceptively on the implications of paranoia among German and British Romantic thinkers.

12. See, for example, Elms' Chapter 1 on 'Freud as Leonardo.'

13. As had Ivan Pavlov before him in the 1870s; see Windholz.

14. See Freud, 'Dostoevsky and Parricide', 1928. J.P. Selten attacked his theory in 'Freud and Dostoevsky' in 1993. More recently, David Burgess has written an entire doctoral thesis to disprove Freud's claim about Dostoevsky's birth trauma and epileptic seizures, affirming instead that they were 'narrative fits created by the demands of the narrative design', a position Dostoevsky's biographer Joseph Frank had earlier promoted.

15. The French psychiatrists slowly began to emphasize persecution over other paranoias around the time Baudelaire published his *Les Fleurs du Mal* in 1857–61. See, for example, Lasègue.

16. For the neurological background see Pfau; Micale and Lerner, Micale, and Gilman et al.

17. This is the explicit juncture where an entire book is needed charting its development; the 'nervous' mood in the Vienna of the young Freud is captured in Morton.

18. For Willis's configuration of 'neurology' see Rousseau (4–22).

19. For the period 1859–75 see Packard.

20. For the English translation see Condillac and Carr.

21. A stunning example of the excitement aroused by nervous anatomy, especially in the brain, is found in letters on the subject by the foremost Romantic neurological illustrator of his era, Charles Bell (see especially 170–1).

22. Leo Braudy had seen connections between the old paranoia and the rise of the novel. The literary connections are further studied in Rothfield.

23. It may appear overly cautious to insist on comparative evidence, but only cultural historians of anatomy are sufficiently suited to assess the validity of the assertion.

24. As cited in Davis (6). It comes from the very late Freud in a letter dated 13 August 1937 to Marie Bonaparte.

25. Some modernists will include Beckett's idiosyncratic forms of nervous laughter in the depiction of his characters. For further discussion of modernism's investments in neurological persecution see Stonebridge, and Kappel. Recently an international consortium of medical scientists has begun to correlate the effects of epilepsy and epileptic types of ailment on artists, composers and writers; for some of their fascinating results see Bogousslavsky and Boller.

26. The two seminal works in America written by their leading historian and literary theorist, respectively: see Hofstadter, and Strauss.

Works cited

Bell, Charles. *Letters of Sir Charles Bell, Selected Chiefly from His Correspondence with G.J. Bell.* London, 1870.

Berrios, German. 'Historical Aspects of Psychoses: Nineteenth-Century Issues.' *British Medical Bulletin* 43 (1987).

Berrios, German, and Roy Porter. *A History of Clinical Psychiatry: The Origin and History of Psychiatric Disorders.* London: Athlone, 1995.

Bogousslavsky, Julien and François Boller. *Neurological Disorders in Famous Artists.* Basel: Karger, 2005.

Braudy, Leo. 'Providence, Paranoia, and the Novel.' *English Literary History* 48 (1981): 619–37.

Burgess, D.F. Narrative Fits: Freud's Essay on Dostoevsky. Ph.D. Thesis. University of Washington, 2000.

Bynum, W.F. 'The Nervous Patient in Eighteenth and Nineteenth-Century Britain: The Psychiatric Origins of British Neurology.' *The Anatomy of Madness.* Ed. Roy Porter, Michael Shepherd and W.F. Bynum. 3 Vols. London: Tavistock, 1985. Vol 1. 89–102.

Carroll, John. *Puritan, Paranoid, Remissive: A Sociology of Modern Culture.* London: Routledge & Kegan Paul, 1977.

Condillac, Etienne, and M.G. Carr. *Condillac's Treatise on the Sensations.* London: The Favil Press, 1930.

Davis, Mike Lee. *Reading the Text That Isn't There: Paranoia in the Nineteenth-Century American Novel.* New York: Routledge, 2005.

De Botton, Alain. *Status Anxiety.* London: Hamish Hamilton, 2004.

Dowbiggin, Ian. 'Delusional Diagnosis? The History of Paranoia as a Disease Concept in the Modern Era.' *History of Psychiatry* 11.41 (2000): 37–69.

Elms, Alan C. *Uncovering Lives: The Uneasy Alliance of Biography and Psychology.* Oxford: Oxford UP, 1994.

Esquirol, Jean. *Des Maladies Mentales.* 2 Vols. Paris: B. Ballière, 1838.

Farrell, John. *Freud's Paranoid Quest: Psychoanalysis and Modern Suspicion.* N.Y.: New York UP, 1996.

——. *Paranoia and Modernity: Cervantes to Rousseau.* Ithaca: Cornell UP, 2006.

Frank, J. *TLS* (10 June 1975): 1–2.

Freud, Sigmund. 'Dostoevsky and Parricide.' 1928. *The Penguin Freud Library.* Vol. 14. Ed. James Strachey and Albert Dickson. Harmondsworth: Penguin, 1990.

——. 'Project for a Scientific Psychology.' 1895. *The Standard Edition of the Complete Psychological Works of Sigmund Freud*. Vol. 1. Trans. James Strachey. London: Vintage, 2001.

Gilman, Sander L., Helen King, George Rousseau, Roy Porter and Elaine Showalter. *Hysteria Beyond Freud*. Berkeley: California UP, 1993.

Gordon, K.G. Madness, Masculinity and the Modern Individual: Paranoia and Gender in Twentieth-Century Narrative. PhD Thesis. McMaster University, 2000.

Habermas, Jürgen. *The Philosophical Discourse of Modernity: Twelve Lectures*. Cambridge, Mass.: MIT Press, 1987.

Hill, John. *Hypochondriasis. A Practical Treatise on the Nature and Cure of That Disorder; Commonly Called the Hyp and Hypo*. London, 1766.

Hofstadter, Richard. *The Paranoid Style in American Politics, and Other Essays*. London: Cape, 1996.

Kappel, Andrew J. 'Psychiatrists, Paranoia, and the Mind of Ezra Pound.' *Literature and Medicine* 4 (1985).

Kraepelin, Emil and A.R. Diefendorf. *Clinical Psychiatry, Abstracted and Adapted from 'Lehrbuch Der Psychiatrie' by A.R. Diefendorf*. New York, 1902.

Kraepelin, Emil and G.M. Robertson. *Manic-Depressive Insanity and Paranoia*. Edinburgh: Livingstone, 1921.

Lasègue, C.E. 'Du délire de persecution.' *Archives Génerales de Médecine* 28 (1852): 129–50.

Lewis, Aubrey J. 'Paranoia and Paranoid: A Historical Perspective.' *Psychological Medicine* 1.1 (1970).

Logan, Peter Melville. *Nerves and Narratives: A Cultural History of Hysteria in Nineteenth-Century British Prose*. Berkeley: California UP, 1997.

Matthews, Henry and John Murray. *The Diary of an Invalid*. London: John Murray, 1820.

Micale, Mark S., ed. *The Mind of Modernism*. Stanford, CA: Stanford UP, 2004.

Micale, Mark S. and Paul Frederick Lerner. *Traumatic Pasts: History, Psychiatry, and Trauma in the Modern Age, 1870–1930*. Cambridge: Cambridge UP, 2001.

Morton, Frederic. *A Nervous Splendor: Vienna, 1888/1889*. London: Wiedenfeld & Nicolson, 1979.

Nabokov, Vladimir. *Pnin*. Melbourne: Heinemann, 1957.

Packard, E.P. *Modern Persecution: Or, Insane Asylums Unveiled*. 1875. New York: Arno, 1973.

Pfau, Thomas. *Romantic Moods: Paranoia, Trauma, and Melancholy, 1790–1840*. Baltimore: Johns Hopkins UP, 2005.

Prichard, James Cowles. *A Treatise on Diseases of the Nervous System*. London, 1822.

Rothfield, Lawrence. *Vital Signs: Medical Realism in Nineteenth-Century Fiction*. Princeton: Princeton UP, 1992.

Rousseau, George S. *Nervous Acts: Essays on Literature, Culture and Sensibility*. Basingstoke: Palgrave, 2004.

Selten, J.P. 'Freud and Dostoevsky.' *Psychoanalytic Review* 80.3 (1993): 441–55.

Smith, Lacey Baldwin. *Treason in Tudor England: Politics and Paranoia*. London: J Cape, 1986.

Steadman, Ralph. *Paranoids: From Socrates to Joan Collins*. London: Harrap, 1986.

Stonebridge, Lyndsey. *The Destructive Element: British Psychoanalysis and Modernism*. Basingstoke: Macmillan, 1998.

Strauss, Leo. *Persecution and the Art of Writing*. Westport, Conn.: Greenwood, 1973.

Tissot, A.D. *Essai Sur Les Maladies des Gens du Monde [Essay on the Maladies of Literary People]*. Lausanne: François Grasset, 1770.

Trotter, David. *Paranoid Modernism: Literary Experiment, Psychosis, and the Professionalization of English Society*. Oxford: Oxford UP, 2001.

Trotter, Thomas. *A View of the Nervous Temperament: Being a Practical Enquiry into the Increasing Prevalence, Prevention and Treatment of Those Diseases Commonly Called Nervous*. London: Longman, Hurst, Rees & Orme, 1807.

Westphal, Karl. 'Uber Die Verrücktheit.' *Allgemeine Zeitschrift für Psychiatrie* 34 (1878).

Windholz, George. 'Pavolv's Conceptualization of Paranoia within the Theory of Higher Nervous Activity.' *History of Psychiatry* 7 (March 1996): 159–66.

7
"Nerve-Vibration": Therapeutic Technologies in the 1880s and 1890s

Shelley Trower

> The first shock of a great earthquake had, just at that period, rent the whole neighbourhood to its centre. Traces of its course were visible on every side. Houses were knocked down; streets broken through and stopped; deep pits and trenches dug in the ground; enormous heaps of earth and clay thrown up; buildings that were undermined and shaking, propped by great beams of wood.... In short, the yet unfinished and unopened Railroad was in progress; and, from the very core of all this dire disorder, trailed smoothly away, upon its mighty course of civilisation and improvement.
>
> Charles Dickens, Dombey and Son, 1846–8, 68

Vibratory movements – shakes, quakes, tremors and convulsions – are often used to describe the impact of modernity in the nineteenth century. Vibration in this context is evocative and productive of massive change, of transformation, destruction and "dire disorder" (Dickens 68). In *Dombey and Son*, Dickens's choice of the most naturally destructive, shattering phenomenon of a 'great earthquake' as a metaphor for the construction of the railway – which was for so many the leading symbol of modernity – challenges any straightforward notion of the railway as the embodiment of progress, the very tracks of progress. While the railway trails "smoothly away, upon its mighty course of civilization and improvement", the "shaking" buildings it leaves behind move only back and forth, to and fro, on the spot (68). Vibration, here, doesn't really get us anywhere. But as a returning motion, a repetitive, rapid motion in one place, it nevertheless brings about collapse.

The most well-known image of how stability and solidity gave way to change and uncertainty is probably the *Communist Manifesto*'s suggestion that as the capitalist bourgeoisie depends on constant change in the means of production, "[a]ll that is solid melts into air" (Marx and Engels 83), which recalls the image of vibration in the prefaces, where competition generated by production is said to be "shaking the very foundations of landed property – large and small" (55, 56). Vibratory movements are not simply metaphors for the condition of modernity, however; the rhetoric of shock corresponds with a persistent concern in the nineteenth century with what vibration is actually doing to things, including the human body, and, above all, the nervous system. From the perspective of the inhabitants of Camberling Town, who suffer the earthquake-like shocks in Dickens's narrative, the tracks may trail "smoothly away" (68), but Dombey's nightmarish journey later in the novel points towards another kind of vibration, which does nothing for his disturbed state of mind. His agitation is intensified by the horrendous rhythm of the carriage wheels on the tracks, beating out an endless reminder of mortality amid a literal modern progress: "the triumphant monster, Death", "the remorseless monster, Death", "the indomitable monster, Death" (298–9). The repetitive vibration of rail travel became a subject of particular concern to medical writers, mostly in its capacity to shake the nerves. This was seen to result in a variety of symptoms, from which the short-tempered and extremely uneasy Dombey appears to suffer, including "anxiety", "irritability" and "restlessness", which were noted in the eight-part report published in the medical journal the *Lancet* in 1862, 'The Influence of Railway Travelling on Public Health' (234). Later in the century, the vibratory motion of bicycle journeys posed a similar problem. The understanding that shocks were conveyed to the nerves in a harmful manner when cycling was supported by the fact that this was a recognized cause of 'railway spine'.[1] George Herschell, for example, pointed out: "It is already a fact well established and known to those who have much experience in the treatment of nerve-troubles that continued longitudinal vibrations communicated to the spinal cord are very injurious, and frequently set up a peculiar condition termed neurasthenia", in light of which he put forward a new diagnostic category: "cyclist's spine" (708).

Railways, bicycles, sewing-machines and the noise of city life were among the factors which appeared in nineteenth-century medical literature to have a damaging impact on the nervous system. Historians of medicine, neurology and psychology, literature and culture have unearthed from this literature a host of conditions, ranging from 'railway spine' and 'neurasthenia' to 'nervous shock' and 'traumatic hysteria', which register modernity as a distinctly shocking, even traumatic age.[2] As the unhealthy impact of modernity has been discussed at some length, and considering that much of the shock of the nineteenth century took a distinctly vibratory form,[3] I want to concentrate here on the mechanics of vibration as a therapeutic method, as a way of attempting to counteract or recover from the damage done to the nervous system.

Vibration as medicine – the theories behind which worked with an understanding that the nerves are in a constant state of vibration – resonated with much older ideas of bodily vibration, which helped to provide a sense of continuity in contrast to the experience of modernity as shattering transformation, as a radical break with the past (as described in Dickens's and Marx and Engels's narratives of the collapsing foundations of the world as we knew it). As a returning motion, which can nevertheless cause collapse, vibration can be taken as a model for historical continuity *and* change. The harmonious and discordant vibrations of musical strings had long provided an image of health and illness, and in the eighteenth century increasingly and more specifically provided a model for nervous activity, as this chapter will observe. In the nineteenth century, the use of musical instruments as well as newer technologies (strings and wires in particular) to describe the nervous system was increasingly combined with how technologies impact on that system, in ways that could be both harmful and healing and painful and pleasurable. In other words, bodies of modernity can be seen as sensitive mechanisms, that are sensitive to vibrating machines: the railway train, the percuteur and the vibratode.

The percuteur

In 1881 Joseph Mortimer Granville wrote a letter to the *Lancet* claiming that railway spine, among a variety of neurasthenic disorders – the causes of exhaustion "ranging from mechanical shock to sexual excess" – can be treated with "nerve-vibration" ('A New Treatment' 671). He had earlier developed this form of treatment as a result of observing women in childbirth, whose pain, he claimed, "might be *interrupted* by appropriate mental and physical methods and appliances" ('Treatment of Pain' 286). For the purpose of interrupting pain, Granville designed the "percuteur", which could deliver "a known number of blows per second" to various parts of the body (287). The first version of this was driven by a spring which operated a lever, which in turn caused an ivory hammer to vibrate. It needed to be wound up. Granville eventually came to prefer the percuteur driven by electricity, largely because the speed of delivery could be sustained and controlled, as he explains in his fullest account of nerve-vibration, *Nerve-Vibration and Excitation as Agents in the Treatment of Functional Disorder and Organic Disease* (24–5, 58).

In the case of railway spine, Granville suggested that travelling causes vibrations to propagate up the spinal column, or vertically, the damaging effects of which can be neutralized if vibrations are sent in the horizontal direction: "I believe the *modus operandi* of the treatment is to counteract the tendency to a lax and flaccid condition of the cord by exciting it to vibrate in a direction *at right angles* to its axis; most of the vibrations which act injuriously on the cord being propagated *in the line* of its axis" ('A New Treatment' 671, emphasis in original). Much as the unhealthy vibrations triggered by rail travel are propagated "*in the line*" of the spine's axis and can be treated by "exciting it to vibrate in a direction *at right angles*" (671), Granville proposed that excessively slow vibrations need be treated with fast rhythmic motions, and *vice versa*. Nerve disorders, in other words, can be caused by vibrations travelling in the wrong direction, being too rapid or too slow, or altogether prevented by disease. The aim of therapy then, was to stimulate the nerves with regular, rhythmical motion, to cause them to be thrown into a healthy state of vibration.

Underlying this practice of nerve-vibration was a theory of nerve function as inherently vibratory, a view that had developed through eighteenth-century Hartleyan theory for example, which held that vibrations in the nerves transmit sensations to the brain.[4] The idea that nerves are hollow, and that animal spirits flow through them, began to give way in the eighteenth century to the understanding that nerves are solid, more like musical strings than tubes.[5] This worked out as a kind of translation into neurological terms of earlier traditions in which the human body and soul were likened to strings. In his earlier work, *Common Mind Troubles* (first published in 1878), Granville seems to have used musical strings as a pre-neurological image for human mind and character, considering 'temper', both 'good and bad', as a matter of tuning or tension without referring specifically to the nerves:

> Temper of mind and character is something akin to the tension of strings in a musical strings in a musical instrument, or the temper of steel. If the stretch be equally distributed, the sounds produced, or the cohesion and elasticity possessed, are well formulated and trustworthy. If there be faults in the quality or character, the vibrations fail to emit a true tone, and the strength is treacherous. (80)

The notion of temperament, as John Hollander points out, "was made to apply almost from the beginning of its linguistic history both to the tuning of strings and to the tempering of various parts of the human soul", which ties in with the Pythagorean idea that both strings and soul should be in tune with the universal harmony of the spheres (25). In antiquity, it was known that the sound of a plucked string will cause another string, tuned to the same pitch, to vibrate in sympathy, which provided an image of how all things can be related harmoniously, from the heavenly spheres to the human soul. Along with good and bad tempers, health and illness from the earliest days of Greek medicine, through the Renaissance and beyond, were described in terms of harmony and discord, dependent on the tuning of a person's strings.[6] As well as in Granville's work, the continuation of this tradition can be seen in the work of George Henry Taylor, who also had interests in vibratory therapy, and will be discussed later in this chapter. Taylor explicitly referred to the Greek tradition, to the god Apollo who symbolized the Pythagorean doctrine of the harmony of the spheres, incorporating in his historical account of medical techniques a passage from the seventeenth-century natural philosopher Francis Bacon, who had conducted experiments with sympathetic vibration:[7]

> The human organization, so delicate and so varied, is like a musical instrument of complicated and exquisite workmanship, and easily loses its harmony. Thus it is with much reason that the poets unite in Apollo the arts of music and of medicine, perceiving that the genius of the two arts is almost identical, and that the proper office of the physician consists in tuning and touching in such a manner the lyre of the human body as that it shall give forth only sweet and harmonious sounds. (qtd. *An Exposition* 65)[8]

Granville's metaphorical use of strings to describe temperament similarly echoed this tradition, while his later work focused on the nervous system. Nerve-vibration, according to Granville's theory, is a kind of tuning process:

> Acute or sharp pain is, I believe, like a high note in music, produced by rapid vibrations [of the nerves], while a dull, heavy, or aching pain resembles a low note or tone, and is caused by comparatively slow vibrations. A slow rate of mechanical vibration will therefore interrupt the rapid nerve-vibration of acute pain, while quick mechanical vibration more readily arrests the slower. ('Treatment of Pain' 287)

Musical strings, and the phenomenon of sympathetic vibration, were increasingly employed in the eighteenth and nineteenth centuries as models for how vibrating nervous systems can be influenced by one another, or by other vibrating things, including the nerves of an audience by a poet or actor,[9] the nerves of travellers by a train, or patients by a percuteur or other therapeutic technologies, as we shall later see. Granville developed the use of musical strings, drawing also on contemporary scientific material, including John Tyndall's work with "sensitive flames" (known to shrink or flare, sing or roar, in response to sound), as a model for the nerves: "if strings or reeds vibrating at the same time, though a short distance apart, will fall into harmony, why is it unlikely that nervous organisms, possessing the same qualities of physical structure, should exhibit a corresponding affinity?" ('Treatment of Pain' 288) Granville was concerned in this way with how the nervous system can be too sensitive, and thus thrown into harmful states of vibration.

Such vibratory states were suffered most frequently by women, according to Granville, because the female nerves were especially sensitive. "The female organism is characterized not inaccurately," he claimed, "though popularly, by the phrase 'finely strung nerves'" (288). The hysterical state to which women were supposedly prone was seen to have its physical basis in a certain type of nervous structure, having moved from its earlier location in the womb.[10] In these cases,

> the nervous elements [are] especially liable to vibrate in their sheaths, and thus exposes them [women] to the risk of falling readily into any rhythm which the will, the physico-mental environment, or perhaps even the proximity of similar structures vibrating in other bodies characterised by the same physical organisation, may suggest. (288)

Vibratory mechanisms provided an image of the female body and were objects to which that body was considered particularly sensitive. As a sensitive mechanism itself, in other words, the female organism seemed to be especially vulnerable to dangerous forms of mechanical vibration, which it was the physician's duty to control. In an environment of dangerous vibrations, Granville warned against "the abuse of nerve-vibration" ('Nerve-Vibration: A Caution' 465). He claimed to have no greater dread than that his percuteur should fall into the hands of patients or their friends. "As well, and more safely, might those gentlemen who are pursuing this course supply their clients with scalpels, galvanism, electricity in its various developments, and hypodermic injections" (465). Although he began his investigations with women, and women were more likely to need treatment due to their finely strung nerves, Granville went on to claim that he did not treat women with his percuteur at all: "I have avoided, and shall continue to avoid, the treatment of women by percussion, simply because I do not want to be hoodwinked, and help to mislead others, by the vagaries of the hysterical state or the characteristic phenomena of mimetic disease" (*Nerve-Vibration* 57). This concern that vibration as therapy might somehow revert or flip back to vibration as the cause of the symptom is further evidence of the ambiguity of the phenomenon. An additional ambiguity is of course produced by the anxiety that, in relieving pain, the percuteur and its like might induce undesirable pleasures.

One possible explanation for Granville's concern about the percuteur, therefore, is that women were using such instruments not so much for medical purposes as to produce pleasurable sensations. The dangers of masturbation as set out by medical professionals in the nineteenth century are well-documented. Rachel Maines's *The Technology of Orgasm* claims that physicians became increasingly concerned to distinguish instruments like the percuteur from the models sold to the public for use at home, being "advised to purchase professional-looking equipment, which could not be confused with consumer models", the latter being used to achieve "titillation of the tissues" (95, 100). But rather than look forward with Maines to the development of the electromechanical vibrator, I have here identified a historical context for Granville's work in which the nervous system is itself described as a musical kind of mechanism, that is vulnerable to other mechanisms, which may disturb or restore a sense of neurological harmony. In response to the uncontrollable excess of vibration experienced on a train for example, which, as well as railway spine, has been considered a cause of sexual excitation,[11] Granville and other physicians can be seen to have developed manageable, medical forms of vibration.

Vibratory energies

The ability to control the frequency of vibration, to provide "a known number of blows per second" (Granville, 'Treatment of Pain' 287), was considered important to all kinds of vibratory therapy. Friedrich Bilz for example, who developed a foot-powered "vibratode" – this being the vibrating end piece that was applied to the human body, and was exchangeable for different shapes – noted that this could produce "up to 3000 vibrations in a minute" (1816). Bilz added that "an expert masseur cannot exceed 350 in the same length of time" (1816). While the new speeds or frequencies of vibration experienced by bicyclists and other travellers were considered detrimental to health, the proponents of vibratory therapy aimed to increase the speed at which their instruments could operate. M.L.H. Arnold Snow's account of a range of medical vibratory devices and practices – including "Head-shakes", "musical vibra-massage", "Charcot's vibratory helmet", "the vibratory handle" and "the vibrating table" – observed some of the speeds at which such instruments could operate, and considered that the "essential elements of a good machine" include its controllability by an operator: "the vibrations of uniform quality should be capable of a range from weak to strong, and the force of impulse and rate, or speed, should be under control (the rate ranging as high as 6500 per minute)" (Snow para. 35). But it was electricity, not only as a way of powering mechanisms such as the percuteur, but in its direct application to the body as a treatment for a range of medical complaints, which could deliver the highest rate of 'oscillations'.

Granville explained that nerve-vibration acts in a different manner to electricity, which introduces a "foreign force" to the system, whereas nerve-vibration can excite a nerve centre to perform its own activity, "helping her [Nature] to do her work with her own forces and in her own way" (*Nerve-Vibration* 12). Electricity was useful, however, for cases of paralysis accompanied by muscle deterioration, when Nature, as it were, has run out of steam.[12] This idea that electricity could impart new force to a system made it possible to imagine that the exhausted might be re-energized or even the dead brought to life. Mary Shelley's *Frankenstein* most famously described this fantasy, while various other science fictions and attempts to revive corpses with electricity proliferated throughout the nineteenth century and beyond.[13] As a therapeutic tool though, according to Granville, this force should vibrate. The constant current is not useful, as it simply uses the nerve as a conductor, it "passes along without throwing it into vibration", while the interrupted current "has the power of throwing any particle of matter, whether living or inorganic, into mechanical vibration" (*Nerve-Vibration* 12). The use of electricity was also recommended as treatment for neurasthenia, epilepsy, ovarian tumours and numerous other complaints, in *The Medical Use of Electricity* by George Beard and Alphonso Rockwell, first published in 1867 and followed by many new editions, remaining the standard text on electric medicine through the 1880s and beyond. However, as later editions note, electricity was also used to execute criminals,[14] and had been the cause of fatal accidents to workmen engaged upon electric-light cables at 1000 volts. Rockwell's experiments, and those of Nikola Tesla (inventor of the alternating current), supported the use of the vibratory, alternating current, as opposed to direct current, as they "seemed to indicate that enormous voltages were harmless to the human body if only they could be made to alternate with sufficient rapidity" (Rockwell 177). Rockwell warned that the harmlessness of high frequency alteration was not yet proved (178), while currents of great frequency could be produced – from "a thousand, or a million, or billion or more oscillations per second" (355).

The vibratory energy thought to power the nervous system was just one among the various forms of energy with which physicists and others were preoccupied in the period, including light and heat, as well as electricity. Medical theories about vibrating mechanisms and electricity usually involved the assumption that these various energies, or at least that electricity and nervous energy, were exchangeable. They could also come into conflict, however. As Armstrong observes, there was "a duality in electricity: seen as duplicating the motive forces of the nervous system and perhaps even the 'spark' of life itself, it was at the same time becoming part of a network of power which transcended the scale of the human body and could kill" (*Modernism* 14). He refers to a range of practitioners for whom "the characteristic nineteenth-century malady of nervous weakness", usually caused by the stresses and strains of modernity, "could be cured with electricity's nerve-like energy", and yet it was electricity that powered much of the technology which contributed to the modern mode of life – involving rapid modes of transport and information flow via telephone and other kinds of wires – that was thought to have exhausted nervous systems in the first place: "demands on nervous energy produced by the complexity of modern life and its speed of locomotion, work, and information flow (as well as factors like the use of stimulants, luxuries, the education of women) resulted in neurasthenia" (15, 17). Electricity seems to problematize the boundaries between exhaustion and exhilaration, or even life and death, as Armstrong suggests. That vibration could provide the "spark" of life for physical matter, and that sound, light, heat, electricity and nervous energy are all forms of vibration, was suggested for example by Maurice Pilgrim, as part of his theoretical justification for the medical use of "vibratory stimulation". Such stimulation was required, according to Pilgrim, when the nerves became 'lethargic' or 'partially unreceptive' to the ethereal stimulus:

It is generally believed and scientifically recognized, that hearing and sight, for example, are the results of rapid vibrations in the universal

ether. Physicists say that the ether here referred to, is a medium fill-
ing all space through which the vibrations of light, radiant heat, and
electric energy are propagated. This medium, whose existence most
authorities now consider to be established, is thought to be more
elastic than any ordinary form of matter and to exist throughout
all known space, even within the densest bodies. Electric and magnetic phenomena are explained as due to strains and pulsations in the
ether. Why may it not be reasonably supposed that the *vis a tergo* –
the vitalizing force behind the anatomical structure of a nerve and
which "makes it go," – is a part of this same operating cause? Nerves
must draw their vitality from some never-failing reservoir, otherwise
the breaks in the continuity of physical life would be of such frequent
occurrence as would very soon complete the extinction of the human
race. (Pilgrim 106–7)

This understanding of life as dependent on the nervous system being
'powered' by, or drawing its vitality from, an external vibratory force,
seemed to make it vulnerable on the one hand to any fluctuations in
that force, which could cause "the extinction of the human race" (107),
and on the other hand to make it the potential beneficiary of "vibratory
stimulation". But he did not see that "the rapidity of the vibrations" –
most instruments ranging from the rate of 1750 to 2500 per minute –
were as important as their "length" (118, 119). "The general 'shake-up'
of the entire body, such as some vibration instruments are constructed
to give", he claimed, "is neither desirable nor productive of satisfactory
results" (120–1).

Vibrations after death

The idea that electricity might have the power to bring inanimate matter, nerve-like strings and wires, or stitched together corpses, to life,
seems to have been familiar at least since *Frankenstein*. Once alive,
according to the theories of spiritualist practitioners of vibratory therapy, it may be that nothing ever dies. The first law of thermodynamics,
that of the conservation of energy, was central to much spiritualist
theory through the second half of the nineteenth century. Energy, it
was argued, is transmitted and transformed, rather than created or
destroyed. Light may pass into heat, for example, beyond the range of
any human capacity to sense it, but it continues to exist. Nervous and
other human kinds of energy are of course also subject to this principle,
according to theorists such as Gustav Theodor Fechner, who was one

of the earliest to explore connections between energy physics and his religious ideas. In *On Life After Death*, first published in 1836, Fechner described activity in the nerves (and brain) as vibratory, and vibration as energy that never dies. He claimed that "vibrations only *seem* to die out, in so far as they spread indefinitely in all directions; or, if dying out for a time, transformed into energy or tension, they are able to begin afresh, in some form or other, in accordance with the law of the conservation of energy" (54). He also compared the radiation of energy out of the brain to that of a stringed instrument, which provides an image of the mechanics of human consciousness that resonates with a long-established tradition, as we have seen:

> However minute and gentle a vibration connected with some conscious movement within our mind may be – and all our mental acts are connected with, and accompanied by, such vibrations of our brain – it cannot vanish without producing continued processes of a similar nature, within ourselves, and, finally, around ourselves, though we are not able to trace them into the outer world. A lyre cannot keep its music for itself; as little can our brain; the music of sounds or of thoughts originates in the lyre or in the brain, but does not stay there – it spreads beyond them. (Fechner 52–3)

Later in the century, the law of energy conservation was used to explain the benefits of vibratory therapy. For George Henry Taylor, inventor of a range of therapeutic, vibratory mechanisms like 'the manipulator', this law explained how massage could heal people:

> Motor energy is transmitted to the vital tissues and the fluids which pervade them in the form of *waves*, whose length and depth are determined by the amount of pressure and extent of motion the hand affords. But it no longer remains motor energy; neither is it lost or even diminished, force or energy being in its nature indestructible. It simply assumes other forms, which, together, are the exact equivalent of that transmitted. (Taylor, *Mechanical Aids* 38)

In other words, massage generates energy within the usually female body – considering the "fearful increase among females of a great variety of nervous disorders" (Taylor, *Diseases of Women* 20) – the beneficial consequences of which, as energy never dies or is "diminished", being "indestructible", start to seem limitless in their potential. Taylor goes on to observe the parallel between the vibrations observed by physicists, and the waves and vibrations produced by his mechanisms, which make it possible to deliver massage at even higher rates, beyond "those natural for the hand of an operator to supply", beyond "the narrow line of human capability" (*Mechanical Aids* 48):

> We have been shown that what to our senses are the far differing phenomena of light, heat, electricity, motion, and other forms of energy, are but variations of *rate* and *degree* of essentially the same thing – differing forms and conditions, modified by material agents, of wave-like or vibratory motions. By the aid of appropriate apparatus, we may also investigate the consequences of vibratory or wave-like motions, at their different rates, when transmitted to the complex components, vital and non-vital of the organism, and learn what we may of the laws governing this procedure and the objects attainable thereby – matters which are as a sealed book to those whose observations have been limited to the degrees and rates of motor energy which emanate from the masseur. (Taylor, *Mechanical Aids* 49–50, emphasis in original)

The idea of energy conservation may have supported Taylor's belief in the powers of the legendary spiritualist medium, Katie Fox, whose career in receiving spirit messages began as a child in 1848 during the internationally reported 'Rochester Rappings'. She was treated by Taylor at various times through the 1870s and 1880s and on hundreds of occasions performed as the medium at séances with the doctor and his wife. *Katie Fox: Epochmaking Medium and the Making of the Fox-Taylor Record*, written by their son, William George Langworthy Taylor, reports that "Katie was, during psychic activity, a dweller at the sanatorium (viii). The energetic principle of spiritualism – of a "logic of continuity" – is set out in the introduction (1–28). Here again vibration appears as a feature of modernity that resists, through its repetition, the change and forward movement of which it is imagined to be a part. Vibration repeatedly brings back the (dead) past.

The various forms of vibratory energy in which the Taylors were interested presumably included that of spiritual frequencies to which Fox was considered highly sensitive. As both a patient and a medium, she was appropriately nervous and delicately feminine, like "a quivering leaf" (Taylor, *Katie Fox* 149). As well as feeling sensitive to Taylor's mechanisms, which could deliver at the rate of "up to 1,600 vibrations per minute" (*An Illustrated Sketch* 18), she may have felt sensitive to the even higher frequencies observed by spiritualists such as William Crookes, in whose experiments she participated during the 1870s ('Notes of An Enquiry' 77–97). Crookes was interested in how the scale of vibration extends beyond the sensory thresholds, ranging from sound, heat and light up to X-rays, an orgasmic "2,305763,009213,693952 per second or even higher", beyond which lies the realm of spiritual vibrations, as he explained in 1897 (*Presidential Addresses* 98).

To imagine the human and especially female body as mechanical, its nerves as strings or wires, may be to see it as sensitive to vibratory intervention – in both its dangerous and beneficial forms. Vibration, it seems, can both harm and heal, cause and cure nervous illnesses, and can also bring inanimate matter to life, and go on after death. Medical practitioners can be seen to have responded to a desire for continuity, to have attempted to heal patients in a modern world. In response to experiences of modernity as shock, or as earthquake-like transformation that produced a break between past, present and future, practitioners of vibratory medicine developed techniques for establishing a sense of relative stability, both by drawing on ancient images of the body as a vibrating set of strings, which seem to survive in newer images of electrically charged 'nerve-wires', and by engaging like spiritualists with the principle of energy conservation. In vibration, modernity survives itself, beyond therapy, in endless movements of neurological return.

Notes

1. See Jane F. Thrailkill's chapter in this collection for an account of a slightly different contemporary meaning of 'railway spine'.
2. See for example Armstrong, 'Two Types of Shock'; and Micale and Lerner.
3. For further discussion of vibratory shock see my 'Upwards of 20,000'.
4. As set out in the 'doctrine of vibrations' in Hartley's *Observations on Man*, first published in 1749.
5. There are many accounts of the movement towards solid nerves, including Clarke's 'The Doctrine of the Hollow Nerve in the Seventeenth and Eighteenth Centuries', and Wightman's 'Wars of Ideas in Neurological Science'. The use of strings as an image for nerves is observed by various historians of medicine and literature, including Finney (156), Roach (94–105), and Miles (49–51).
6. For an account of the continued use in the seventeenth century of musical strings as images for the body and soul in relation to the cosmos see for example Finney.
7. See Gouk for an informative account of Bacon's experiments in relation to both earlier natural magic traditions and the development of the science of acoustics (see especially 159–70).

8. Taylor takes this quotation from Bacon's *The Advancement of Learning: Book 2* (first published in 1605).
9. See for example Roach; Goring.
10. Hysteria's original location in the womb is observed by Jane F. Thrailkill in this collection. Thrailkill's work can be seen to take a different but related route to this, with its emphasis on a new kind of neurological discourse as it applies to the male body.
11. For examples of physicians who saw railway vibrations as a cause of sexual excitation see Maines's commentary (90–1), and for psychoanalytic perspectives see Schivelbusch (77–9).
12. See also Granville, 'Nerve-Vibration as a Therapeutic Agent' 951.
13. For an account of this see Armstrong, *Modernism, Technology and the Body* 13–41.
14. The first chapter of Armstrong's *Modernism, Technology and the Body* begins by describing two executions, one by hanging after which friends tried to revive the corpse with electricity, and the other in the electric chair (13–4), and goes on to discuss how electricity is used both to kill and to heal, which is to be considered further below.

Works cited

Armstrong, Tim. *Modernism, Technology and the Body*. Cambridge: Cambridge UP, 1998.

——. 'Two Types of Shock in Modernity.' *Critical Quarterly* 42.1 (2000): 61–73.

Beard, George and Alphonso Rockwell. *Medical and Surgical Uses of Electricity: New Edition*. London, 1891.

Bilz, Friedrich Eduard. *The New Natural Method of Healing*. London: A. Bilz, 1898.

Clarke, Edwin. 'The Doctrine of the Hollow Nerve in the Seventeenth and Eighteenth Centuries.' *Medicine, Science, and Culture*. Ed. Lloyd Stevenson and Robert Multhauf. Baltimore, Maryland: Johns Hopkins UP, 1968.

Crookes, William. 'Notes of an Enquiry into the Phenomena called Spiritual During the Years 1870–1873'. *The Quarterly Journal of Science* 11 (1874): 77–97.

——. *Presidential Addresses to the Society for Psychical Research 1882–1911*. Glasgow: Robert Maclehose, 1912: 86–103.

Dickens, Charles. *Dombey and Son*. Oxford: Oxford UP, 1974.

Fechner, Gustav. *On Life after Death*. London: Searle & Rivington, 1882.

Finney, Gretchen. *Musical Backgrounds for English Literature: 1580–1650*. New Brunswick, New Jersey: Rutgers UP, 1961.

Goring, Paul. *The Rhetoric of Sensibility in Eighteenth-Century Culture*. Cambridge: Cambridge UP, 2005.

Gouk, Penelope. *Music, Science, and Natural Magic in the Seventeenth Century*. London: Warburg Institute, 1999.

Granville, Joseph Mortimer. *Common Mind Troubles*. 1878. Montana: Kessinger, 2003.

——. 'A New Treatment for Certain Forms of Neurasthenia Spinalis.' *Lancet*. 15 October 1881. 671.

——. 'Nerve-Vibration as a Therapeutic Agent.' *Lancet*. 10 June 1882. 949–51.

——. 'Nerve-Vibration: A Caution.' *Lancet*. 16 September 1882. 465.

———. *Nerve-Vibration and Excitation as Agents in the Treatment of Functional Disorder and Organic Disease*. London: J. & A. Churchill, 1883.

———. 'Treatment of Pain by Mechanical Vibrations.' *Lancet*. 19 February 1881. 286–8.

Herschell, George. 'Bicycle Riding and Perineal Pressure: Their Effect on the Young.' *Lancet*. 18 October 1884. 708.

Hollander, John. *The Untuning of the Sky: Ideas of Music in English Poetry, 1500–1700*. New Jersey: Princeton UP, 1961.

'The Influence of Railway Travelling upon Public Health.' *Lancet*. 1 March 1862. 231–5.

Maines, Rachel. *The Technology of Orgasm: "Hysteria," the Vibrator, and Women's Sexual Satisfaction*. Baltimore, Maryland: Johns Hopkins UP, 1999.

Marx, Karl and Frederick Engels. *The Communist Manifesto*. London: Penguin, 1967.

Micale, Mark and Paul Lerner. *Traumatic Pasts: History, Psychiatry, and Trauma in the Modern Age, 1870–1930*. Cambridge: Cambridge UP, 2001.

Miles, Robert. *Anne Radcliffe: The Great Enchantress*. Manchester: Manchester UP, 1995.

Pilgrim, Maurice F. *Vibratory Stimulation, its Theory and Application in the Treatment of Disease*. New York: Metropolitan Publishing, 1903.

Roach, Joseph. *The Player's Passion: Studies in the Science of Acting*. Ann Arbor: Michigan UP, 1993.

Schivelbusch, Wolfgang. *The Railway Journey: The Industrialization of Time and Space in the 19th Century*. Berkeley: California UP, 1986.

Snow, Mary Lydia Hastings Arnold. *Mechanical Vibration*. 2005. Early American Manual Therapy Version 5.0. 28 March 2007. <http://www.meridianinstitute. com/eamt/files/snow/mvch2.htm>.

Taylor, George Henry. *An Exposition of the Swedish Movement-Cure*. New York: Fowler & Wells, 1860.

Taylor, George Henry. *An Illustrated Sketch of the Movement Cure: Its Principles, Methods and Effects*. New York, 1866.

———. *Diseases of Women: Their Causes, Prevention, and Radical Cure*. Philadelphia: Geo. MacLean, 1871.

———. *Mechanical Aids in the Treatment of Chronic Forms of Disease*. New York: George W. Rodgers, 1893.

Taylor, W. G. Langworthy. *Katie Fox: Epochmaking Medium and the Making of the Fox-Taylor Record*. New York: G. P. Putman, 1933.

Trower, Shelley. 'Upwards of 20,000: Extrasensory Quantities of Railway Shock.' *The Senses and Society* 3.2 (July 2008): 153–67.

Wightman, W. 'Wars of Ideas in Neurological Science – from Willis to Bichat and from Locke to Condillac.' *The History and Philosophy of Knowledge of the Brain and its Functions*. Ed. Frederick Noel Lawrence Poynter. Oxford: Blackwell, 1958.

8

From Daniel Paul Schreber through the *Dr. Phil Family*: Modernity, Neurology and the Cult of the Case Study Superstar

Michael Angelo Tata

> The adventure of our childhood no longer finds expression in *'le bon petit Henri'*, but in the misfortunes of 'little Hans'. The *Romance of the Rose* is written today by Mary Barnes; in the place of Lancelot, we have Judge Schreber.
>
> Michel Foucault, *Discipline and Punish*, 1975, 194

Psychic singularities

Alongside modernity, with its emphasis on the rational (Descartes), the secular (Weber) and the bureaucratic (Bell), a notorious antihero explodes into popularity, a creature I term the 'case study celebrity' – perhaps the strangest development within the history of neurology since Santiago Ramón y Cajal first stained the neuron using the Golgi process in order to perceive it as a totality in 1887. From 'neuron' to 'nervous' psyche, this neurological quantum finds its most Romantic form in the interpretations of coincident disciplines psychoanalysis and sexology, Victorian practices in which nervous disturbance deviates from scientific method in fostering a culture of the diseased and a poetics of the neural. These famous Others are truly Romantic in the sense of nineteenth-century British Romanticism, embodying a cerebral and corporeal alterity that runs counter to the reason-dominated subject of Enlightenment thinking. An exemplary individual whose fame derives from its moral and aesthetic distance from a shaky standard of bourgeois 'salubriousness' of the sort described by Sigmund Freud in his *Civilization and Its Discontents*, the eminent freak – eminent because he or she is both case study protagonist and antihero of rationalism – haunts modern logic and rationality with the spectre of mental and bodily collapse. Via the work of psychoanalysis (Freud, Jung and Jones), sexology (Krafft-Ebing) and natural science (Darwin), modern thought redefines the human being as a contingent entity whose health or illness can only be accidental. Poised at the brink of an all-dissolving counter-modernity and functioning as portal for superstition, ritual, magic and the strange gravity of the irrational, the case study protagonist embodies this very fragility. The notorious and decisive break between Freud and Jung in 1912 testifies to an analogous instability, demonstrating psychoanalysis's own scientific crisis: Is it science *per se*, or merely a literary venture? While Freud would like to posit his nascent discipline as rigorously scientific, others, like Michel de Certeau, have later read him as a literary critic or armchair anthropologist. Helen Vendler has even read him as something of a poet, coining the category 'The Freudian Lyric' as an important contemporary form. Clearly, Freud lies somewhere between the human sciences (*Geisteswissenschaften*) and the natural sciences (*Naturwissenschaften*), a positioning to which the strange and marvellous case of Daniel Paul Schreber, himself sandwiched between competing diagnoses of neuroanatomy (Paul Flechsig, Guido Weber) and psychoanalysis (Freud), speaks.[1]

Shining bright against the dull, matte background of a compulsory socioeconomic normalcy, the neurological Other captivates the imagination with its utter relinquishment to repertoires of desire unthinkable for the standard citizen inhabiting and mastering nation-state and colonial empire. As elucidated by Michel Foucault theoretically in his *The Will to Knowledge*, as well as materially in memoir trouvé *Herculine Barbin: Being the Recently Discovered Memoirs of a Nineteenth-Century French Hermaphrodite*, the modern world takes as its supreme project the rigid fixing of the body's limits and borders in the genesis of docility and domestication. For Herculine Barbin, the oscillation between the poles of female and male necessitates juridical intervention, as when in 1860 the French government orders that she live, dress and conduct the business of her life as a man ('Abel'), barring her from the female (and lesbian) identity around which her world had been structured. The fixing of borders such as Herculine's gender, however, does not normalize anything, instead engendering a fantastic realm of chimaeric human monsters whose inclinations and proclivities perform the amusing work of *grotesquerie*; in other words, these demons and aberrations from a natural order cavort and frolic with ghoulish glee for a rapt audience frozen between fright and pleasure (for example, the John Ruskin of *The Stones of Venice*). As with Raphael's or Bosch's grotesques, monstrous epitomes of lust and appetite proliferate in playful visual displays for the eye of their Freudian beholders, who witness their shenanigans from the safe and neutralizing distance of the parlour, drawing room, lecture hall, museum, vitrine or library.[2] For these spectators, beholding the grotesque Other, such as Barbin-as-literary-artefact or textual relic, defines the safe aesthetic pleasure of witnessing the bizarre from afar.

Of this special set of monsters, the peculiar case of German judge Daniel Paul Schreber looms large. Along with fellow Freudian travellers Dora (1905), the Rat Man (1909) and the Wolf Man (1918), he delineates an apparent boundary between health and illness at the same time that he destabilizes any difference between the two zones. As with his fellow celebrities, he is an allegorical figure representing one particular human vice: for Schreber, paranoia, for Dora, hysteria and for Little Hans, phobia. Similar to characters in a Medieval morality play who perfectly embody one aspect of the human experience, such as 'lust' or 'sloth', Schreber achieves synonymity with 'paranoia', a term posing the danger of 'exhausting' him in the philosophical sense (i.e. summing him up without remainder), as it does with regard to Freud's analysis of his passive homosexuality and paraphrenia. Outside psychoanalysis proper, other stars multiply, many famous for 'being the first' to personify a unique malady, as with Ludwig Binswanger's 1944–5 case study of Ellen West, subsequently the "world's first" anorexic.[3] Hence while a Medieval or Renaissance saint, like St. Catherine of Siena, might starve herself into oblivion, she still cannot own anorexia, nor can it own her, since she, unlike West, is not consumed by the renunciation of hunger. Binswanger comprehends the cultural value of a case study heroine like Ellen West: "On the basis of the life-history, her specific name loses its function of a mere verbal label for a human individuality – as that of this unique time-space-determined individual – and takes on the meaning of an eponym (*fama*)" (267). Taken for a melancholic by Kraepelin (257), diagnosed by Bleuler as a schizophrenic (266), and finally discharged from Wolfgang Binswanger's Kreuzlingen Sanatorium only to leave and commit suicide, West leaped into history by starving *herself* into oblivion long before Karen Carpenter dieted herself out of existence or Oprah Winfrey publicly collapsed from diet exhaustion. Like these and other famous deviants, Schreber uses his very existence to throw into chaos the notion of a 'normal' human being, thus being exhausted by his condition at the same time that by and through it his humanity crystallizes: if even a prominent judicial expert with enviable property, family and career can find himself slipping away into a miasma of hallucinations, what hope is there for those of us hanging by a thread already?

Even going so far as to erect a phantasmic, competing neurology of his own design, Schreber echoes his era's *Zeitgeist* through his bizarre participation in the science of stimulation, enervation, vibration and impulse transmission: he, too, is a neurologist, taking his own body as experimental object and using it to generate and test hypotheses. For example, Schreber theorizes about the relation that obtains among nerves, gender and sexual arousal. In Schreber's system, nerves of Voluptuousness, or *Wollustnerven*, are primarily the property of women, who seem to have these spread across their bodies evenly in the manner of, to pervert an image from contemporary physics, cosmic radiation dispersed throughout a galaxy. Subjected to the pernicious influence of rays in general, Schreber reverberates with the re-imaging and re-imagining of materiality that the physics of his and previous eras had brought to fruition, a world in which matter and energy are revealed to be interdependent and, to some extent, interchangeable: not to the $E = mc^2$ extreme, yet of a part with Sir Isaac Newton's experiments with rays of lights and prisms in the *Opticks* of 1704, or Wilhelm Conrad Röntgen's discovery of electromagnetic radiation in 1895.[4] Schreber's eventual unmanning (*Entmannung*) causes him to develop lust-nerves in particular. While the male body does contain signs that perhaps it, too, once contained pleasure nerves (e.g. the vestigiality of male nipples), it has long since been rid of them (Schreber 61). Finally as a result the dispersal of *Wollustnerven*, women are predominantly bisexual: "Further, the souls knew that male voluptuousness is stimulated by the sight of female nudes, but on the contrary female voluptuousness to a very much lesser extent if at all by the sight of male nudes, while female nudes stimulate *both* sexes equally" (155). Theories such as these reveal Schreber as both patient and doctor alike, offering a heretical diagnosis and aetiology of his malady.

In his *Memoirs of My Nervous Illness* (1903), Schreber recounts in florid, vivid prose his transformation from juridical expert to belching, prattling, transgendered trollop, therein documenting the very anxieties confronted by the modern creature of reason as first systematized by Descartes in his *Meditations on First Philosophy*. Schreber's illness is thus utterly vital as blueprint or 'secret history' of modernity, occurring not as psychotic hiatus with a progressive present, but rather as supremely "of the moment" and symptomatic of a historical now.[5] As Eric Santner demonstrates in his *My Own Private Germany: Daniel Paul Schreber's Secret History of Modernity*, this historical now was a pre-Hitler Germany on the verge of immersion in the totalitarian milieu – a *Deutschland* replicated in miniature in the disciplinary theories of Schreber's father, Moritz, to whom Schreber himself refers:

> Souls knew very well that a man lies on his side in bed, a woman on her back (as the 'succumbing part,' considered from the point of view of sexual intercourse). I myself, who in earlier life never gave it a thought have only learned this from the souls. From what I read in for instance my father's MEDICAL INDOOR GYMNASTICS (23rd Edition, p. 102), physicians themselves do not seem to be informed about it. (155)

That Schreber's historical now involves the familial 'now' of his being the son of a famous theoretician of childhood disciplinary practices and devices, Moritz Schreber, sets up a microcosm/macrocosm dynamic, with a fledgling pan-Germanism of clear-cut bloodlines and adherence to tradition informing and reciprocally being informed by stern paternalism and blind devotion at a nuclear level. Later analyzed by Freud as historical residuum in his foundational *Psychological Notes Upon an Autobiographical Account of a Case of Paranoia (Dementia Paranoides)* (1911), Schreber's account of his supposed victimization by a mortifying divine force blowing in from pre-modernity forms the basis for a theory of the rational human being's assault by perceived endangering forces atavistically forcing him or her to exit the loop of modernity and contemporaneousness for a vengeful, mystical and untamed past. He is also important to the modern taxonomy of mental illness; for Freud, it is essential to determine whether 'paranoia' is a separate entity from schizophrenia, or if the two are fused, as Kraepelin had suggested, and Schreber is just the man to illustrate his case because of the discreteness of paranoia as diagnostic category (151–4). Through Schreber, *wahnbildungsarbeit*, or 'the work of delusion-formation', enters the lexicon.

Keeping good company with the necrophiles, gender dysphorics and sadomasochists populating Krafft-Ebing's *Psychopathia Sexualis* (1886), Judge Schreber exposes a vibrant counternarrative within modernity, one which persists, despite any and all attempts at institutionalization, abjection or eradication. If the normative narrative of modernity involves a reason centred in Descartes' cogito, then the counternarrative authored by Schreber and others is the story of the irrational and unreasonable, as well as of the dispersed self (a self not neatly contained within the boundaries posed by a cogitating substance, or *res cogitans*).

Himself a product of neurology, Schreber emerges from within its folds as a modern synaptic oddity and paragon of masculinity: neurology, psychoanalysis and poetry all mandate his existence, ensuring his survival for a future committed to the allure and glamour of dysfunction. In his person, the future of both the German state and humanity in general are implicated. Schreber is manhood in the extreme – an odd state of affairs, given his transformation into God's consort. Hence although in his daily life he had been unable to reproduce and had instead chosen to adopt his daughter Fridoline, in his fantasy life he possesses the position and means to repopulate an entire planet with his progeny – but only after he has developed *Wollustnerven*. Oddly enough, Schreber becomes most masculine when he becomes most feminine, using his transformed body to bear the divine race which will inherit an earth stripped bare of human life after its cleaving zygotes have taken root in his uterus.

Jung's identification of a Collective Unconscious in his essay 'The Structure of the Psyche' in 1927 makes it possible that even the most Cartesian cogito may awaken to discover a non-personal Archetype from a long-gone enchanted era reasserting its presence via the logic of anachronism, as he discussed anecdotally in his 'Instinct and the Unconscious' of 1919.[6] In Jung's example, a man who, convinced that the sun is blowing wind through a tube and striking his eye with the resulting jet of air, blinks repeatedly, neither hallucinates nor dissociates: he has merely succumbed to a resurgent Mithratic memory of the sun's penis which has sedimented within a group psyche and returned to the present, where, unrecognized, it creates the illusion of psychosis (36). Similarly, for existential psychology (primarily R.D. Laing and Thomas Szasz), there is no schizophrenia, only the social construct of the deviant or, as phrased by Deleuze and Guattari, *le schizo*; though not due to any past historical apparition reasserting itself, here the schizo's madness stems from the social need to categorize certain behaviour as pathological, and in essence disappears, since pathology becomes mere social construct and hence loses its pathological quality. For theories such as these, there is no possibility of celebrity, no opportunity for the everyday human being to leap out of banality in the creation of psychological notoriety, since there is technically no insanity, only a societal need to institutionalize those whose world-views are different and dangerous. To visit another competing theory of derangement, John Paul Sartre's Existential Psychoanalysis put forth in his *Being and Nothingness* founds itself upon the radical split between two realms or orders of being, the inhuman in-itself (*l'être en soi*) and the human for-itself (*l'être pour soi*). For Sartre, Existential Psychoanalysis is possible when Freudian determinism (drive, ego) is eradicated so that the for-itself might cor-rectly regard itself as 'unthinglike' and stop living as if it were either an in-itself (thing) or in-itself–for-itself (the for-itself taking itself for a thing, such as happens with personality, type or ego). In Sartre's sys-tem, there is also no chance of psychic celebrity, since mental illness is not something which might be owned or even embodied: rather, it becomes a mis-ontologization remedied only through the ego's dissolu-tion. The existence of the career psychotic or neurotic reveals a strange liaison between the Freudian psychic map and the celebrity of the out-law. Reduced to an *en soi*, and coinciding absolutely with his illness, a case study protagonist like Schreber owns his nervous impulses so completely that they become a species of property, radical even by the seventeenth-century's standards.

Along with figures like Ellen West, who owns anorexia, Dora, who owns hysteria, or the *Dr. Phil Family*, which owns the neurosis *du jour* (everything from teen pregnancy to cult brainwashing, depending on the season), Schreber carries his illness with him everywhere he goes, tracing out an arc of fame which more staid members of the bourgeoisie can consume as cathartic corrective for their own sobriety.[7] With Schre-ber, medical representation and self-representation emerge jointly and coevally, but only on a map with blurred boundaries and a very fuzzy key: still, his exterior representation as psychotic and delusional by the psychiatric and psychoanalytic community and his interior represen-tation as awash in the pleasures of divine prurience converge, making Schreber's version of Schreber a poeticized take on clinical language in general. Against all odds, this Schreber sets himself up as author not only of his own psychological destiny, but also of a delicious and sala-cious memoir, carrying on a legal battle regarding their publication at the same time that he conducts intergalactic conversations and finds his body swapping genders. Dr. Guido Weber's analysis of the *Memoirs* and their fitness for publication reads like a recipe for literary illustriousness:

> When one looks at the content of his writings, and takes into con-sideration the abundance of indiscretions relating to himself and others contained in them, the unembarrassed detailing of the most doubtful and aesthetically impossible situations and events, the use of the most offensive vulgar words, etc., one finds it quite incom-prehensible that a man tactful and otherwise of fine feeling could propose an action which would compromise him so severely in the eyes of the public, were not his whole attitude to life pathologi-cal, and he unable to see things in their proper perspective, and if

the tremendous overvaluation of his own person caused by lack of insight into his illness had not clouded his appreciation and the limits imposed on man by society (Addendum B, 'Asylum and District Medical Officer's Report' 347–8).

The back cover of the 2000 edition of Schreber's *Memoirs*, designed specifically to entice readers rather than to repel them or prevent them from existing in the first place, as had Guido Weber's dismissal, puts it more succinctly, revealing their commodity power one century later: "His book is perhaps the most revealing dispatch ever received from the far side of madness." Of course the notion of ownership I am using with regard to Schreber and paranoia is problematized by the fact that, in the end, it is not Schreber who, via the *Memoirs*, owns paranoia, but paranoia which owns Schreber – and Freud who, via interpretation, owns this ownership. There are certainly other paranoid schizophrenics, but there is only one Daniel Paul Schreber, originator of the discourse surrounding paranoia, interpreted into notoriety by Freud, and leaving behind a ghastly account of bodily workings and cosmic transgressions.

In league with the case study's most extreme limit personality, the serial killer, the case study maverick garners domestic and global attention alike only within a culture in which the commodity can flourish without interruption or impediment: capitalism. As discussed by Richard Tithecott, the serial killer represents the secret perverseness of capitalism, displaying a mania for accumulation according to which parataxes of shopping and hoarding are displaced into the realm of bodies (instead of amassing a handbag collection, or, if he is Ed Gein, constructing one from pelvic fragments, the serial killer collects corpses, the more, the better). This dirty interior Schreber, too, shares, demonstrating that, according to the logic of capital, even a disequilibrium within neurotransmitter levels is something which might be privately owned and culturally mobilized as a selling point: hence the legal wranglings surrounding Schreber's publication of his *Memoirs*, which could only see the light of day after he had been released from tutelage by the Royal Superior Country Court of Dresden in 1902. With regard to this volume, Schreber's ownership of them is a matter for other judges to decide. Their conclusion relegates Schreber's words to Schreber, returning his memoir and the rights regarding their publication to him: "Therefore in considering the Appeal which has been lodged this must lead to the tutelage inflicted on plaintiff being rescinded without entering into new evidence by witnesses offered by him" (Schreber 440). Marketed by Freud as advertisement for psychoanalysis, Schreber and other paradigms of psychic disturbance populating Freud's pristine terrain not so much prove a theory (here, that of the *wahnbildungsarbeit*) as much as launch a new discourse (Freudian psychoanalysis) and cultural icon (Freud). It is thus no accident that, of all competing psychoanalyses, it is Freud's and not Jung's (too mystic), Adler's (too alpha-male) or Lacan's (too philosophical) which captures the public imagination, since it is specifically Freud's which best provides for private ownership of maladies, thereby making possible a cosmography of world-famous lunatics and sexual deviants with whom it becomes imperative for literate and refined human beings to familiarize themselves. This heterological tradition imports the *demimonde* into the Academy, where it stands as corrective for philosophy's adherence to a dry Kantian epistemology and aesthetics, Critical Theory's untiring focus on commodity evils, or psychology's naïve belief in perfectibility.

God's plaything

If *Senätsprasident* Daniel Paul Schreber is anything, he is a visionary neurologist, one for whom vibrating nerve fibres exist as religious and cosmic conduit. Himself an intense neurological study, he is far too educated a human being to abide by the prescriptions of, for example, a Kraepelin, whose theory of hallucinations Schreber finds grossly materialistic – so much so that he openly disagrees with him, launching a more comprehensive theory: "If psychiatry is not flatly to deny everything supernatural and thus rumble with both feet into the camp of naked materialism, it will have to recognize the possibility that occasionally the phenomena under discussion may be connected with real happenings, which simply cannot be brushed aside with the catchword 'hallucinations' " (84). Fundamentally empiricist, Schreber lays out his *Memoirs* methodically, and with attention to thematic organization (God's miracles follow an organic clustering, united temporally along a timeline and physically in various bodily regions). From the start, Schreber is interested in providing a philosophical structure to his story – this despite the fact that some find his ruminations to constitute only an "unconscious parody of the preoccupations of philosophy" (Dinnage xix). In Schreber's structure, a minimum register of foundational axioms provides support for a more hefty network of secondary elaborations; in fact, within this arrangement, signs of Schreber's mental illness do not pop up until the twenty-second page of text, at which point Schreber casually, even credibly, interjects: "I will at present only mention the fact that the sun has for years spoken with me in human words and thereby reveals herself as a living being or as the organ of a still higher being behind her" (22). As laid out in the introductory chapters to his *Memoirs*, several immutable facts and conditions make possible all the mayhem that follows. One example of such an axiom is the postulate that, without exception, God does not possess the capability of understanding the live human being: "*A fundamental misunderstanding* obtained however, which has since run like a red thread through my life. It is based upon the fact *that, within the Order of the World, God did not really understand the living human being* and had no need to understand him, because, according to the Order of the World, He dealt only with corpses" (62). For God, the human being can only be known as expired block of flesh – hence his continual misapprehensions of Schreber, and the tortures inflicted upon him via divine miracle. Atomic, a statement as this crucial one achieves fulfilment later on in the text, when Schreber's bowels come under scrutiny. Convinced that God does not find him sufficiently intelligent to empty his own cloaca, Schreber uses defecation to prove to God that, indeed, his mind is in working order:

> The question 'Why do you not then sh...' is followed by the capital answer 'Because I am somehow stupid.' The pen almost resists writing down the fantastic nonsense that God in his blindness and lack of knowledge of human nature in fact goes so far that He assumes a human being could exist who – something every animal is capable of doing – cannot sh...for sheer stupidity. (206)

As vile as the image of an "eliminating" Schreber experiencing the "soul-voluptuousness" of defecation may be, its importance lies in the fact that an intricate neurological and phenomenological theory is being proven by way of bodily behaviour that can only be compulsive (206).

Since God cannot comprehend either human thought or the human organ of thought, brain activity functions as a perpetual source of ignorance and divine cruelty. Thus for Schreber the absolute limit to his own intellective stream in the 'thinking-nothing-thought', or *Nichtsdenkungsgedanke*, emerges as a particular source of anxiety. Recalling the idea of intentionality championed by Brentano and later challenged by Sartre – i.e. the notion that consciousness must always be conscious of something – this delicate and intuitive concept does more than merely posit oases of non-thought or *négatités*, cerebral voids where thinking nothing is possible.[8] For within the detailed system built by Schreber – the first of its sort to be carefully written down with an eye to its systematicity and hence the original masterpiece of *wahnbildungsar-beit* – any lapse in active thinking proves the divine hypothesis that Schreber is himself some sort of idiot whose speech can only echo that of birds senselessly twittering away with meaningless fragments of human speech. When human language, as embodied by Schreber, meets the holy language of *Gründsprache*, or "Basic Language", cataclysm and pain can be the only outcomes. Essentially a language of reversal (euphemism) and aporia (the sentence fragment), the speech inflicted upon him by mystical entities like Flechsig, von W or any of the "little men" who populate his ontologically bustling world drive him mad by causing him constant semantic and grammatical labour.[9] Having to de-euphemize euphemisms in order to determine the true malignancy of their content, while simultaneously substituting the missing words from half-uttered sentences, Schreber finds no respite from the work of translation: compared to him, a contemporary Schreberian *memoiriste* like Anne Heche ('Celestia', as she renames herself in *Call Me Crazy*, the narrative of her own descent into madness) will have it easy.[10] Overloading his nerves, constantly having to decode everything uttered in *Gründsprache* causes Schreber's own nerves to resonate, thereby launching an unchecked internal dialogue of *Nervensprache*, or 'Nerve Language':

> Normally under normal (in consonance with the Order of the World) conditions, use of this *nerve-language* depends only on the will of the person whose nerves are concerned; no human being as such can force another to use this nerve-language. In my case, however, since my nervous illness took the above-mentioned critical turn, my nerves have been set in motion *from without* incessantly and without any respite (55).

Identified by Schreber as another axiom, the idea that nerves possess their own channel and means of communication provides clues as to Schreber's own illness. Because *Gründsprache* is essentially an inverted and inchoate language, it necessitates the constant motion of *Nervensprache*. Consequently, there is no 'thinking-nothing-thought', with the rare exception of musical activity (the piano, the ocarina), or with the violent intrusion of 'interferences' afflicting his intermittent equanimity – for example, the Bellowing Miracle, Schreber's *stoppage* of choice:

> The advent of the bellowing-miracle occurs uncontrollably when my muscles serving the processes of respiration are set in motion by the

lower God (Ariman) in such a way that I am forced to emit bellowing noises, unless I try very hard to suppress them; sometimes this
bellowing recurs so frequently and so quickly that it becomes almost
unbearable and at night makes it impossible to remain in bed. (188)

Interferences such as these chronic attacks of bellowing prevent Schreber from having a moment's rest, placing him on pins and needles
perpetually and forcing his musculature into a non-stop tonic state
which can only cause fatigue and debilitation.

The system in which Schreber finds himself imbricated, and which
takes him as "Eternal Jew" and star, is cacophonous and verbally disruptive.[11] In addition to the voices of the "Forecourts of Heaven" or
"Tested Souls", spoken in a cold and exasperating *Gründsprache*, there
is also another parasitic language of roteness and mindless repetition:
the ornithological language with which various birds prattle on.[12] In
this language, nonsensical substitutions cause perennial confusion –
for example, "Santiago" and "Cathargo" are absolutely equivalent, as
are "Chinesenthum" and "Jesum Christum" and "Ariman" and "Ack-
ermann", among other words and expressions (192–3). Though not all
birds speak this dead language – pigeons and chickens do not seem to
trouble him – those that do torture him with the constant spectre of his
own lapse into automatic speech:

> To the miraculously created birds belong all fairly *fast flying* birds,
> particularly all singing birds, swallows, sparrows, crows, etc., *of these
> species of birds I have never once during these years seen a single spec-
> imen which did not speak.* . . . On the other hand the pigeons in the
> court of the Asylum do *not* speak, neither as far as I have observed
> a canary kept in the servants' quarters, nor the chicken, geese and
> ducks which I have seen both from my window on the plots of land
> lying below the Asylum, and on the two mentioned excursions in
> villages through which I passed; I must therefore presume that these
> were simply natural birds (196).

Mistaking Schreber for one of these birds, God must indefatigably test
him in order to ensure that the one human being who will repopulate the earth with his progeny is not some kind of avian robot
or twittering machine out of a Paul Klee canvas.[13] The 'delusions of
grandeur' which will become a paranoid schizophrenic staple find their
earliest expression in Schreber's own notion that, somehow, the world
has ended during his institutionalization. Apocalyptic, Schreber finds

himself placed in the unique position of having been chosen to repopulate a razed earth with 'the Schreber race'. Finally beyond the threats of Catholicism, pan-Slavism and Judaism, which haunt him throughout the text, Schreber now ponders the awesome responsibility of inheriting an entire planet – a true task, given that he will also have to contend with inimical religious forces as well, although these will be rephrased in the language of radiation:

> Thus while still in Flechsig's Asylum I became acquainted with rays – that is to say complexes of blessed human souls merged into higher entities – belonging to the old Judaism ("Jehovah rays"), the old Persians ("Zoroaster rays") and the old Germans ("Thor and Odin rays") among which certainly not a single soul remained with any awareness of the name under which it had belonged to one or other of these peoples thousands of years ago. (30)

During his transfer from Flechsig's institute to 'The Devil's Kitchen', the streets of Leipzig appear in lurid theatricality. All these buildings are merely stage props installed to provide the illusion that the human species continues to dominate a spinning globe from which it has in actuality been expelled:

> I was inclined to take everything I saw for a miracle. Accordingly I did not know whether to take the streets of Leipzig through which I traveled as only theater props, perhaps in the fashion in which Prince Potemkin is said to have put them up for Empress Catherine II of Russia during her travels through the desolate country, so as to give her the impression of a flourishing countryside. (100)

For Schreber, urban sprawl has been reconstituted as *mise-en-scène*, his life a dramaturgical production starring himself as diva and heroine.

Yet even if this vision proves incorrect, and Schreber has not been selected as the last receptacle of the *Homo sapiens* genetic treasure, then other means of creation remain not only viable, but inevitable. Hence Schreber entertains fantasies of metempsychosis, transmigrating into a specific chain of bodies: a Hyperborean woman, a Jesuit, a Burgomaster, "an Alsatian girl who had to defend her honor against a victorious French officer", a prince from the Mongol Horde (88). These cross-cultural and transgendered entities represent critical transmutations marking Schreber as undifferentiated flesh or, if a more contemporary metaphor may be endured, 'stem cell', a man in possession of organs which deteriorate, become other organs, or switch functionalities, and whose identity transcends the constraints of space and time. Perverting Haeckel's Biogenetic Law of 1866, Schreber demonstrates an incredible degree of pluripotency, his cells seeming to have retained an ability to re-draw the course of his own development. Even if none of these transformations come to pass, there is always the possibility of intergalactic travel, as when two brothers of Cassiopeia initiate a conversation with Schreber, or when he finds himself living for a time on Martian moon Phobos.

Central to Schreber's importance to future generations, though, is his emasculation, or *Entmannung*. This seed which will germinate into the full-fledged belief that his body has biologically changed from male to female comes early in the narrative, just before Schreber's illness consumed him. Prior to institutionalization, Schreber is seized by the intrusive thought of what it would be like to be a woman on the other side of copulation: "Furthermore, one morning while still in bed (whether still half asleep or already awake I cannot remember), I had a feeling which, thinking about it later when fully awake, struck me as highly peculiar. It was the idea that it really must be rather pleasant to be a woman succumbing to intercourse" (46). Up until this point, Schreber had never imagined what it would be like to be 'woman', but from all outward appearances the surfacing of this thought sets in motion his concomitant decline as judge and rise as cerebral architect. Soon after, the very God who cannot comprehend any live human entity, let alone one on the verge of discovering his secrets, takes Schreber as his consort, catapulting him into a moral crisis. Oddly enough, what bothers Schreber is not the withering away of his own male genitalia, or even the blossoming of 'nerves of voluptuousness' throughout his entire body, but only the use to which his new body will be put. For Schreber, these are two potential uses of his body, one eschatological and holy, the other ephemeral and unsavoury. Whether the sex he has with God is procreative or merely pleasurable torments Schreber; hence the oscillation of Ariman's two uttered phrases "voluptuousness has become God-fearing" and "Excite yourself sexually" – expressions denoting Madonna and whore states (252). What he can tolerate, even desire, is becoming a uterus; what he cannot stomach is the chance that God might once again misunderstand him and view him as no more than vulva.

Ultimately, Schreber the scientist prevails, fabricating meaning from the miracles under which he suffers by way of an inventive neurology according to which a proprietary voluptuousness belonging only to women is made accessible to him. An M-to-F who retains his male character throughout, and who seems to encounter the most voluptuousness when he expels faeces, not when his newly developed genitalia are put to use, Schreber dreams of an orgasm spread across his body while never achieving it; as such, he prefigures even Lacan, for whom *jouissance* transcends mere "oases of pleasure" for a wholly sentient bodily surface area.[14] Why God never defiles him, or why there is not even the presence of a sexless conception as occurs in Christianity, are questions never answered: Schreber worries about the sexual uses to which his new body might be put for nothing. While various men, of course, do magically appear inside Schreber, as when, in the most famous instance, a gaggle of Benedictine monks takes up residence inside his head, along with Flechsig phantasms and a wandering von W, who functions as a constant reminder of the chronic urge to masturbate, none appear to be sex offenders:

> The Jesuit Father S. in Dresden, the Ordinary Archbishop in Prague, the Cathedral Dean Moufang, the Cardinals Rampolla, Galimberti and Casati, the Pope himself who was the leader of a particular 'scorching ray,' finally numerous monks and nuns; on one occasion 240 Benedictine Monks under the leadership of a Father whose name sounded like Starkiewicz, suddenly moved into my head to perish therein. (57)

Male Matroyoshka doll, Schreber is XY through and through, despite the transmigratory eventuality of becoming a Hyperborean woman or sprouting pleasure neurons. Fusing neurology with both personal history as well as German, cosmological and religious history, Schreber is a battlefield on which various factions play. Subject to miracles and counter-miracles in dazzling succession, Daniel Paul Scrheber does, indeed, give birth, but not to a race of giants. Rather, his child is the fundamental worldview of the paranoiac, a tight, tidy clockwork whose greatest miracle is the genesis of a self that is the whole of the cosmos.

Psychotainment

If post-structuralism is right in its disavowal of originality and truth, then every reading is a misreading in one way or another; consequently, it appears impossible to get testy with Freud for his uses of Schreber, just as it would from this perspective be gauche to fault him for getting Leonardo da Vinci's history wrong in *Leonardo da Vinci and a*

Memory of His Childhood. Here, despite the fact that so much of this work hangs on Freud's mistranslation of the word *nibbio* (which he interprets as 'vulture' rather than 'kite'), the fantastic system that Freud himself is able to construct based upon something so minor as a detail from a dream makes even fractals seem simple.[15] One good *wahnbildungsarbeit* deserves another, and so it is with Schreber that two of the world's greatest system-builders come into contact on a madman's heated page, warm from the glow of miraculated embers. Although Freud will ignore much of Schreber's experience, naturally focusing on the imagined scene of emasculation central to Schreber's encounter with his strange and sadistic God, he is still correct in his identification of Schreber as one for whom delusions can sediment into glorious structures more compact ('condensed') and regulated than a Ptolemaic solar system, yet still vibrant and surprising, nonetheless.[16] Since for Freud the human narrative rests upon a sexual dynamic informing all aspects of daily life, it is impossible that Schreber escape such categorization. And so Freud will fix his gaze on Schreber's homosexual attachment to his evil genius therapist, averting his eyes from both the sensual excesses of Schreber's text in and of itself, as well as the material conditions of Schreber's imprisonment and the effect these facts of his existence exert upon Schreber's own mental development – for example, the connection between his isolation, as detailed by his maps of "God's Nerve Centers", and the fantasy that the human race has followed Cuvier's hypothesis and gone extinct, or the impact that being isolated in the women's wing at Flechsig's asylum could have had upon his gender identity (72).

The fact that delusions can form intricate structures following idiosyncratic yet well defined rules seems to be Schreber's gift to posterity. For a world committed to projects of self-fashioning, as has been shown by the work of Michel Foucault regarding the care of self, Stephen Greenblatt on self-construction in the Renaissance, and Charles Taylor on 'sources of the self' within modernity, a self as exotic as Schreber's will function as a role model of sorts. He has certainly been the inspiration behind Deleuze and Guattari's concept of the *corps sans organs*, his own philosophical fame emerging as the greatest miracle of all, proof that his confessionality has transcended even Freud's attempts at containment. Recalling other inordinately complicated systems, such as that developed by William Blake, for whom an entire pantheon of personalities and types is suspended on axes marked 'innocence' and 'experience' in prophetical works like *Jerusalem* and *Visions of the Daughters of Albion*, the self as fashioned by Schreber and re-fashioned by Freud enters the public domain as world-historical and exemplary. While of course Freud could never have predicted what would happen once, a century later, the camera's lens turned itself towards Bedlam, he is in a sense responsible for inventing the very system by which the fame of lunacy could be launched and appreciated. With the advent of contemporary phenomena like Oprah Winfrey product and protégé *Dr. Phil*, who has graced the airwaves of American television since September 2002, textual pyrotechnics, such as those housed in Schreber's glistening prose, give way to visual histrionics, as audiences across the globe gasp in horror when a father takes a lie detector test which will prove whether he has been fondling his daughter's 'penis', or shake their heads in disbelief when a misandrous shopaholic accused of buying her teenage children porn reads off items on a recent credit card bill. Scenes such as these should come as no surprise to creatures of modernity (even if that modernity is 'post-'). For centuries, the psychotic has been entertaining – perhaps even trephined cave people were the court jesters of their time, Paleolithic images awaiting discovery by the work of future spelunkers. While it would take Romanticism proper to turn the monster, blood-sucker or the demented into a viable counter-hero, such a development could never have taken place without the installation of subject-centred reason at the core of a culture committed to progress. Where we have finally arrived, terms like 'repression', 'transference' or 'countertransference' are less meaningful than what can happen on a stage when human nerves are stripped raw of even their myelin sheaths and left to fester under harsh studio lights. Kleptomaniacs, sexual predators and binge eaters sell Nesquick and Geico insurance, placing themselves exactly where they belong: next to cartoon bunnies and talking geckos. Were he alive, Schreber would be there, too, tooting on his ocarina and rebounding with lambent voluptuousness.

Notes

1. In *Heterologies*, Michel de Certeau outlines other psychoanalytic crises within the work of Freud. For example, although he acts the part of anthropologist in a work such as *Totem and Taboo*, he performs no exotic field work, deriving his complex theory of a totemic root for phobic behaviour, as well as his story of the murder of the primal father, from the reports of other anthropologists. Similarly, he derives his theory of paranoia from a literary analysis of Schreber's memoirs, not from direct observance of the man. Regarding the relationship between the *Geisteswissenschaften* (human sciences) and *Naturwissenschaften* (natural sciences), see Rorty. See Vendler for a discussion of the Freudian Lyric.

2. I take my sense of the grotesque from Wolfgang Kayser's *The Grotesque in Art and Literature*. In this work, Kayser meticulously traces the development of this important aesthetic category from Raphael onward.

3. Paradoxically, Binswanger concurs with Bleuler regarding Ellen's schizophrenia (*schizophrenia simplex*), using Janet's case study of Nadia to illustrate why exactly neither woman suffers from anorexia (331–7). Still, Binswanger provides intricate details of Ellen's personal battle with consumption, including her poetry as evidence of her split between wispy ethereality and Jewish terrestriality, thereby generating a very detailed anorexic artefact. His analysis of West's hunger as "existential craving," or *Süchtigkeit*, represents the *Dasein* of starvation (345–7).

4. Schreber mentions Röntgen rays in particular, connecting them with memory, history and materiality:

> During the first years of my illness it would in my opinion have been an easy matter by a thorough examination of my body with the help of medical instruments and above all with Roentgen-Rays (not then discovered) to demonstrate the most obvious changes in my body, particularly the injuries to my internal organs which in other human beings would have been fatal. If it were possible to make a photographic record of the events in my head, of the lambent movements of the *rays coming from the horizon*, sometimes very slowly, sometimes – when from a tremendous distance – incredibly swiftly, then the observer would definitely lose all doubt about my intercourse with God. (303)

5. For Santner, Schreber's narrative is both a secret history of modernity and a secret history of Nazism. As Dinnage elucidates in her introduction to Schreber's memoir, the 'household totalitarianism' of Daniel Paul's father, Moritz Schreber, helps to clarify Schreber's delusions, as so many of these involve sensations of restraint and loss of control and as such represent a dissolution of Moritz's famous system of discipline. They are, in essence, a symbolic and sexualized reconstitution of his restraining device, the Schreber *Geradehalter*. Freud, too, notices the inherent Nazism of Schreber's delusions, noting that even in Schreber's heaven, the darker, Semitic God, Ariman, is inferior to the lighter Persian-Aryan God Ormuzd.

6. One problem with Jung's collective Unconscious is its apparent Lamarckian dimension. In Darwin's theory of evolution (but not necessarily competing theories, such as Bergson's), there is no way for present modifications of plant or animal matter to be transmitted to future generations. Consequently, Jung's psychic model presents the problem of inheritance: how can a cultural memory be stored and passed along via DNA? The paradigm of the Archetype (e.g. the Mandala, or the young dying god) reveals a basic structure to human consciousness, an organization which compels the psyche to think certain patterns (e.g. a world-dissolving flood). 'Categorical' and 'Noematic', these structures infuse human thought with a recognizable and transmissible form.

7. One critical difference between Dr. Phil's case studies and Freud's is that Dr. Phil's do not launch new illnesses, but rather redefine existing problems

according to modern interpretations. What is not new about Dr. Phil is the diagnosis – for example, anorexia. His novelty stems from the ways in which his television program and publications personalize a type of pre-existing psychological condition: on his show, the professional anorexic serves her runaway husband with divorce papers.

8. For Sartre, *négatités* are 'little pools of nothingness' which interfuse daily human life – for example, the absence of a friend, the subsumption of a detail by 'ground' in the emergence of 'figure'. As such, they run counter to Heidegger's conception of nothingness as the void along which human existence is stretched.

9. In Schreber's cosmography, the figures of therapists Flechsig and von Weber loom large. Since it is Flechsig, the therapist treating him for his initial breakdown, whom Schreber accuses of the very "Soul Murder" which sets in motion his unmanning, it is Flechsig who becomes the tormentor. At points, there are even a Superior and Middle Flechsig (113); these effect the miraculous "Tying-to Rays" and "Tying-to-Celestial-Bodies" by which Schreber's own body is stretched across the heavens and connected to the universe. In addition, Schreber also achieves nerve contact with Flechsig's ancestor Daniel Fürchtegott Flechsig, whose presence reveals a rift between the Flechsig and Schreber families. "Little men" are entities ("Tested Souls") who have been pared down to "one single nerve" (75).

10. In her very Schreberian *Call Me Crazy*, Anne Heche metamorphoses into Celestia, a quasi-religious interplanetary traveller charged with writing a new Bible – curiously enough, in leather books she purchases at a Christopher Street sex shop. Laying out her new language, Heche clues her reader in to a revolutionary language not unlike Schreber's *Gründsprache*.

11. Whether Ice Age (59) or syphilitic outbreak (79) has destroyed the human race, Schreber remains behind as Eternal Jew (60) to repopulate the earth. Following Georges Cuvier's theory of catastrophism, he posits himself as answer to mass extinction. Later, he is even given a "Jew's stomach", this along with other physical tortures (a dissolved penis, a miracled coccyx [151]).

12. Like St. Teresa of Avila, Schreber lays out a planar view of heaven. After expired humans become "Tested Souls" and are duly subjected to their respective ordeals, they return to occupy the Forecourts of Heaven, or *Verhöfe des Himmels*. Next to these are the Anterior Realms of God. In front of these, the Posterior Realms of God appear, these divided into the realms of Ormuzd and Ariman. Ormuzd orchestrates Schreber's *Entmannung*, while Ariman becomes a source of Zoroaster Rays.

13. I can imagine no better visual analogue of Schreber's pesky birds than Paul Klee's *Twittering Machine* (1922). This painting depicts a merchandized bird gizmo twittering away rotely. While Klee bears no reference to Schreber, the painting does synchronicitously provide a perfect visual representation of Schreber's ornithological terror.

14. For Lacan, the body originally experienced orgasm wholly, and only postcastration divided the torso into "oases of pleasure" (a penis, a clitoris). Passing from *jouissance* to mere *plaisir*, the body differentiates into zones of sexual stimulation and zones of sexual indifference – which, incidentally, become sexualized through the detours of the fetish.

15. For *nibbio* means 'kite', not 'vulture', and with that, the structure of specu-
    lation that Freud built on that miraculous bird, the vulture, collapses. But
    the theory retains its interest – for example, the idea that scientific inquiry
    begins with "the sexual researches of children," or that the empiricist is a
    brooder by nature. As the editors of the *Standard Edition* rightly note, what
    remains intact is Freud's "detailed construction of Leonardo's emotional life
    from his earliest years, the account of the conflict between his artistic and
    his scientific impulses, the deep analysis of scientific theory" (Gay xxiii).
16. Towards the end of his exposition of Schreber's condition, Freud does come
    clean, admitting that his theory of paranoia was developed prior to his hav-
    ing read Schreber. This, taken along with the fact that he finds it more
    productive to examine "exciting causes" than Schreber's own words, demon-
    strates the value Schreber has for Freud and for psychoanalysis in general (II:
    Attempts at Interpretation, 110, 118). It thus should come as no surprise that
    'castration', 'father-complex' and 'passive homosexuality' should be where
    Schreber leads him: Freud was already there.

Works cited

Binswanger, Ludwig. 'The Case of Ellen West.' *Existence: A New Dimension in Psy-
    chiatry and Psychology.* Ed. Rollo May, Ernest Angel and Henri F. Ellenberger.
    New York: Simon & Schuster, 1958.
de Certeau, Michel. 'The Freudian Novel: History and Literature.' 'Mystic Speech.'
    'The Institution of Rot.' *Heterologies: Discourses on the Other.* Trans. Brian
    Massumi. Minneapolis: University of Minnesota, 1986.
Dinnage, Rosemary. Introduction. *Memoirs of My Mental Illness.* By Daniel Paul
    Schreber. Trans. Ida Macalpine and Richard A. Hunter. New York: New York
    Review of Books, 2000.
Dr. Phil. 'The Lie Detector Tests Part 2' (7 November 2006); 'The Lie Detector
    Tests Part 1' (6 November 2006); 'Shocking Accusations' (3 November 2006);
    'Alex Turns 18' (follow-up with 'The Anorexic Bride') (16 May 2006); 'Shopping
    Intervention: The Aftermath' (26 January 2006); 'A Shopping Intervention'
    (4 November 2005); 'Extreme Food Obsessions' (3 November 2005). CBS.
Foucault, Michel. *Discipline and Punish: The Birth of the Prison.* Trans. Alan
    Sheridan. New York: Vintage, 1991.
——. *Herculine Barbin: Being the Memoirs of a Nineteenth-Century French
    Hermaphrodite.* New York: Pantheon, 1980.
——. *The Will to Knowledge.* 1976. Vol. 1 of *The History of Sexuality.* 3 Vols.
    London: Penguin, 1998.
Freud, Sigmund. *The Psychotic Dr. Schreber. Three Case Histories.* Ed. Philip Riefe.
    NY: Touchstone, 1996.
——. *Leonardo da Vinci and a Memory of His Childhood.* Trans. Alan Tyson. New
    York: W.W. Norton, 1989.
——. *Civilization and Its Discontents.* Trans. James Strachey. New York: W.W.
    Norton, 1961.
Gay, Peter. Introduction. *Leonardo da Vinci and a Memory of His Childhood.* By
    Sigmund Freud. Trans. Alan Tyson. New York: W.W. Norton, 1989.
Heche, Anne. *Call Me Crazy: A Memoir.* NY: Scribner, 2001.

Jung, Carl Gustav. 'The Structure of the Psyche.' 'Instinct and the Unconscious.' 'Individual Dream Symbolism in Relation to Alchemy.' *The Portable Jung.* Ed. Joseph Campbell. NY: Penguin, 1976.

Kayser, Wolfgang. *The Grotesque in Art and Literature.* 1957. Trans. Ulrich Weisstein. Bloomington: Indiana UP, 1963.

'New Diet Threatens Her Life: Oprah Collapses.' *The National Enquirer.* 16 October 2006.

Rorty, Richard. *Philosophy and the Mirror of Nature.* Princeton: Princeton UP, 1980.

Santner, Eric. *My Own Private Germany: Daniel Paul Schreber's Secret History of Modernity.* Princeton: Princeton UP, 1996.

Sartre, Jean-Paul. *Being and Nothingness.* Trans. Hazel E. Barnes. New York: Washington Square Press, 1984.

Schreber, Daniel Paul. *Memoirs of My Mental Illness.* Trans. Ida Macalpine and Richard A. Hunter; New York: New York Review of Books, 2000.

Tithecott, Richard. *Of Men and Monsters: Jeffrey Dahmer and the Construction of the Serial Killer.* Milwaukee: Wisconsin UP, 1999.

Vendler, Helen. *The Given and the Made: Strategies of Poetic Redefinition.* Cambridge: Harvard UP, 1995.

9

"I guess I'm just nervous, then": Neuropathology and Edith Wharton's Exploration of Interior Geographies

Vike Martina Plock

> I wonder if it isn't the way of all nervous illnesses to oscillate in that clock-like way? I'm always si e no [Italian: yea-and-nay] when I am tired – a gray day and then a – well, relatively pink one! – But I wish the affirmative days wd, in your case, join hands and exclude the others.
>
> Edith Wharton to Bernard Berenson, 4 January 1911,
> *The Letters of Edith Wharton*, 232

Edith Wharton was no stranger to nervousness, neurological ailments and, in particular, neurasthenia. Although the long-established claim that she was, in 1898, subjected to Silas Weir Mitchell's notorious rest cure has recently been challenged by Shari Benstock (93–4),[1] Wharton's biography is, through the nervous disorders of two of her closest male companions, her husband Teddy Wharton and her literary friend Henry James, punctuated by the experience of neuropathology. In fact, it was "dear H.J." (15 October 1907, *Letters* 116), who, in 1910, took the rest cure himself while Wharton expertly described his case as one "for a neurologist" (19 March 1910, *Letters* 202). Wharton, it seems, knew what she was talking about. During considerable parts of her marriage and especially from 1909 to her divorce in early 1913, she had to observe how her husband's "sweetness of temper and boyish enjoyment of life struggled long against the creeping darkness of neurasthenia", from which, according to the opinion of "all the neurologists we consulted", "there could be no recovery" (*A Backward Glance* 326). In an unusual reversal of gender dynamics, it was the men in Wharton's life who, in

184

the first decades of the twentieth century, succumbed to supposedly feminine nervous pathologies, whereas the 'masculine', rational and cool-headed Wharton was charged with supervising their recoveries.

It is perhaps not surprising, then, to learn that Wharton, the *grande dame* of American literature and novelist of manners, also had, in addition to these very personal confrontations with neurology, a life-long, ardent interest in contemporary sciences. Her "commonplace book lists the definitions of a large array of sciences and scientific philosophies" (Lewis 108),[2] while her reading list included the evolutionary works of Charles Darwin, Herbert Spencer, Thomas Huxley and Ernst Haeckel (Lewis 56; Singley *passim*). One particular branch of dominant contemporary science, however, did not sit well with Wharton: "the psychological-pietistical juggling" of "William o' the wisp James" (21 February 1906; 24 March 1910, *Letters* 101–2, 205). And although Wharton's aversion to Henry James's elder brother might well have been personally motivated, Robin Peel suggests that it was mainly the "anti-intellectualism" and the "strand of antirationalism...with the celebration of the sensation, the vision, the epiphany" (Peel 112) in William James's, at times, anti-materialist psychology and especially his fusion of neurological and psychological research that offended Wharton's belief in logic, reason and rationality. Furthermore, for Wharton William James was, through his famous 'stream of thought' or 'stream of consciousness' model, tightly connected to the literary aesthetics of the modernists.[3] Given the additional impressionist composition of the Jamesian text,[4] Wharton, who famously sneered at the formal experimentation of the next generation of writers, "seeing it as regressive and preoccupied with the abnormal and the pathological" (Peel 112), expectedly renounced James's monumental *The Principles of Psychology* (1890) on both philosophical and aesthetic grounds.

In spite of these obvious misgivings Wharton's passion for science and her intellectual tussle with William James vitally informed her literary output in the turbulent years before her divorce. After publishing the book of short stories *Tales of Men and Ghosts* (1910), and while working on *The Custom of the Country* (1913), Wharton wrote the short novella, *Ethan Frome* (1911), a text which was "written in the only quiet time I've had in the last distracted two years" (23 September 1911, *Letters* 259) and which explicitly foregrounds Wharton's interest in neurophysiology and neurology. Unconventionally for Wharton, *Ethan Frome* is set in rural New England, in the appropriately named fictional village of Starkfield, a town marked by its "sluggish pulse" and said to be "rich in pathological instances" (*Ethan Frome* 6, 48). In what is certainly Wharton's grimmest tale, characters are "nervous as a rat" (8), have "troubles" or "complications" (73), are "just nervous" (63) and constantly "seek the advice of some new doctor" (42).

Even the setting of *Ethan Frome* – a snowy, isolated New England panorama – has a clinical character: it resembles a "pathological chart" (48) and provides a suitable atmospheric background for the thwarted, adulterous love affair between the protagonist, Ethan Frome, and his wife's cousin Mattie Silver. When Zeena, Ethan's sickly wife, decides to cast Mattie out of the house and the lovers see the fulfilment of their sexual and emotional desires frustrated, the couple attempt suicide by simulating a sled accident. This, however, leaves them both disfigured and semi-paralyzed. Moreover, the resulting "disease of the spine" gives the formerly young and vivacious Mattie a "bright witch-like stare" (115) and turns her into the double of Ethan's nagging wife Zeena. Ethan's life with the dissatisfied women drags on. More disturbingly, Wharton's novella opens 24 years after the fatal sled accident, forcefully pointing to the forlornness that marks the dissonant triadic relationship in the Frome household.

This gloomy account of social and emotional paralysis and disappointment is couched into a frame narrative, told by a scientific observer: an urban engineer with an interest in biochemistry. But although the engineer narrator repeatedly insists on getting "the missing facts of Ethan Frome's story" (7), Wharton's narrative is "profoundly connected with the problem of an interior story that cannot be told" (Hutchinson 221). And it is precisely on grounds of the text's interpretative elusiveness and its frame narrative, a stylistic anomaly for Wharton, that *Ethan Frome* has often been compared to Conrad's 1899 modernist novella *Heart of Darkness* (Cahir 20–3; Fryer 184). However, the framed core narrative is only the most potent example of the novella's interest in interiors.[5] Indeed, if Wharton's unexpected recourse to unusual stylistic devices establishes a potential and unexpected link between her novella and the formal experiments of her much scorned and taunted modernist contemporaries, I want to show that her experimentation with narrative form can be related to her interest in contemporary neurological and neuroscientific debates. As we shall see, *Ethan Frome* systematically interrogates the creation of interiority concepts in turn-of-the-century psychological and neurological research – the attempt to develop accurate and definite maps of human interiors.

In this Whartonian examination of neurophysiological and neuropathological profiles, one particular scholar and his writing took centre stage: William James. In writing *Ethan Frome*, Wharton rejected a Jamesian union of metaphysics and science by representing the interior geographies of her characters in purely physiological imagery. Most significant in this context is the text's preference for the anatomical terms 'brain' and 'head' over the more metaphysical expressions 'mind', 'spirit' or 'consciousness': "As he strode along through the snow the sense of such meanings glowed in his brain" (19); "through his tingling veins and tired brain only one sensation throbbed" (38); "little phrases she had spoken ran through his head" (113).[6] Because of her interest in scientific materialism Wharton willingly considered neurophysiological research because it was tangible and exact but rejected emerging psychological theories such as James's as imprecise and speculative. This is why she openly challenged some elements of James's research, but accepted other aspects of the nineteenth-century neurological and neuroscientific debate.

Reading *Ethan Frome* in reference to contemporary neurology and neuroscience can therefore help to analyze Wharton's complex stance towards turn-of-the-century science that preferred neurology and neuroscience over psychology. But we can also relate Wharton's debate with contemporary sciences to the development of her literary aesthetics. Tracing *Ethan Frome*'s reliance on neuroscientific and neurological imagery shows, for instance, how thoroughly the novella adopts beliefs in biological determinism – a concept central to naturalist aesthetic practices. Yet paradoxically, while Wharton's text undoubtedly appropriates a naturalist perspective that is spurred by contemporary neuroscientific research, it simultaneously questions the analytical conclusions of modern sciences – theories that were instrumental in developing this naturalist viewpoint in the first place. Wharton's belief in her own naturalist stance therefore started to slip at precisely the moment when she expressed scepticism about the epistemological course of modern science. It is at this point that *Ethan Fome*'s modernist inflections come into play.

Cords of communication and electric currents

In the time period during which Wharton was writing her major works, neuroscientists developed elaborate concepts of human interiors. In particular, Camillo Golgi's silver chrome staining technique, introduced in 1873, made nerve cells and the detailed organization of the central nervous system visible under the microscope. But what neuroscientists such as Golgi, Santiago Ramón y Cajal and Wilhelm Waldeyer saw when looking through their microscope was, first and foremost, a complexly

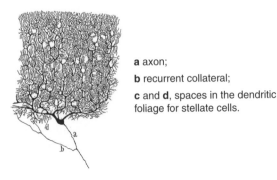

a axon;

b recurrent collateral;

c and **d**, spaces in the dendritic foliage for stellate cells.

Figure 4   A human Purkinje cell stained using the Golgi method. From Santiago Ramón y Cajal, *Histology of the Nervous System of Man and Vertebrates*, I: 54
Source: By permission of Oxford University Press, Inc.

structured and presumably impenetrable web of nerve cells and fibres (see Figure 4) – a highly sensitive and receptive apparatus that was nevertheless capable of multifarious and clearly coordinated neuronal operations that, in turn, resulted in synchronized motor reactions. And it was, no doubt, the nervous system's alleged mechanistic composition and its ostensibly automatic functioning that invited the widespread comparison with modern technological devices. For instance, in his extensive 1904 textbook *Histology of the Nervous System of Man and Vertebrates*, the Spanish neuroscientist Santiago Ramón y Cajal insisted on this suggestive analogy:

> The organic machinery of the nervous system is indeed comparable ... to such ingenious machines as the music box, camera, phonograph, and so on, where simple finger pressure on a spring, or the weight of an inserted coin, activates the whole series of coordinated events the machine was designed to carry out. (I: 16)

However, whereas the nervous system was associated with modern technology in general, a favourite comparison linked it more explicitly to the telegraph. Unsurprisingly, since nerve paths were conventionally seen as "cords of communication" (Ferrier 2), nineteenth-century physicians such as Joseph Mortimer Granville repeatedly noted the resemblance between nerves and telegraph wires:

> These nerves of sensation are simply lines of communication; and if they happen to be cut anywhere in their course – as telegraph wires have been cut in times of war — it is impossible to tell whether the

impressions or message of sense or pain received through them comes from the sense-organ at the extremity or from the seat of injury *en route*. (3–4)

In his attempt to illustrate the nervous system's complex functions for his readers, Granville consciously employed this popular comparison. But in underlining the mechanistic elements in human nervous physiology, he also suggested that its organization was, like modern telecommunication devices, very susceptible to potential disturbances.

Modes of communication and their potential disruption are central to *Ethan Frome*, and the novella openly trades on the metaphorical potential suggested by common association between the nervous system and modern communication devices. The novella's narrator first observes Ethan Frome, the object of his analytical interest, at the local post-office window, where Frome picks up "a copy of the *Bettsbridge Eagle*" or the occasional "envelope addressed to Mrs. Zenobia — or Mrs. Zeena — Frome, and usually bearing conspicuously in the upper left-hand corner the address of some manufacturer of patent medicine and the name of his specific" (4). Most significantly, however, communication lines are very often cut or interrupted in Wharton's novella. For instance, the geographical isolation of Starkfield with its "slow and infrequent" trains (42) and its blocked railroads is experienced by its eponymous protagonist as "being in an exhausted receiver" (18–19).[7] And when a local horse epidemic "spread to the ... Starkfield stables" (9) the narrator finds himself temporarily without means of transport.

This geographical seclusion is mirrored by the breakdown of intellectual and emotional associations. Forced to care for his ailing mother, whose "voice was seldom heard" "after her 'trouble'", Ethan suffers from the silence that "had deepened about him year by year" (46). Only the arrival of his cousin Zeena, charged with nursing Ethan's mother through the last stages of her illness, interrupts "the mortal silence of his long imprisonment" (46). However, shortly after their marriage Zeena "too fell silent" (48), and Wharton's text gives a vivid description of the broken and intermitted flow of communication between husband and wife: "When she spoke it was only to complain, and to complain of things not in his power to remedy; and to check a tendency to impatient retort he had first formed the habit of not answering her, and finally of thinking of other things while she talked." (48) As an immediate result, Zeena, "within a year of their marriage", develops "the 'sickliness' which had since made her notable" (48). And although Zeena immerses herself in books on kidney troubles (96) and clearly seems to suffer from heartburn (83), Wharton text's unmistakably suggests that her ailments are neurological: Zeena's "shooting pains" (42) that run "down to [her] ankles" (43) indicate neuralgia. But this neurological ailment – neuralgia – in resembling the pathological configurations in hysteria cases is of course highly appropriate in the narrative's context. Zeena's actual complaints have been displaced by finding a new and moreover a physiological referent. Hence with the translation of Zeena's verbal protest into somatic signs we can already register Wharton's interest in converting emotionality, and in Zeena's case emotional tension, into plainly detectable physiological markers.

Moreover, neuralgia itself was understood, in turn-of-the-century neurological research, as a displacement disease: the site and the source of the pain are not identical. Indeed, Zeena might feel pain in her ankles but the organic cause of her 'troubles' could well be, as Joseph Mortimer Granville explained in his 1884 textbook *Nerves and Nerve Troubles*, "in the stomach or some other part of the alimentary canal". Granville commented: "This remoteness of, and apparent want of connection between, cause and effect is a source of great embarrassment in attempting to relieve this variety of pain" (45). Whereas the locality of the actual pain and its organic root stand in a metonymic relationship, it is the nerve pathways that establish these pathological associations and produce the respective physiological symptoms. As Granville indicated, in employing another comparison with modern technology, neuralgic symptoms index disturbed and abnormal relations in the nervous system "as though a telegraphic system intended for the service of a community should by some freak of force begin to send about messages or indications of its own" (42). In Zeena's case the somatic symptoms that her body displays are referential in a doubled sense. Her collapsing biological 'telegraphic system' indicates the interrupted or misdirected flow of both nervous *and* social communication.

Complaints such as Zeena's were the happy hunting ground for turn-of-the-century neurologists. And given the accepted and widely used analogy that nerve experts established between the human nervous system and modern technological machinery, it is hardly surprising that electricity was widely used as a trope to render the workings of the nervous system understandable. But whereas some neurologists and neuroscientists simply conceptualized the flow of nervous energy by stressing its affinity with electrical currents, others, such as Francis Howard Humphris, argued for the structural equivalence of the two:

I now wish to point out and trace the similarity between our nerve force and electricity.... It is needless to discuss the analogy between electric phenomena produced by electric fishes and the currents produced by the human tissues, for it is admitted now that all tissues of living organisms are capable of producing electric currents. (79–80)

Support for advocating this overlap of the biological and technological machine, that based its working on electric force, came from the controversial vivisectionist research of neurologists, including the Scottish physician David Ferrier. In his attempt to draw up exact cortical maps and to "prove that specific regions of the cortex controlled particular zones of the body" (Otis 28) Ferrier, from 1873 onwards, inflicted lesions and electrical stimulation to the cerebral cortex of dogs and other laboratory animals. The results of those experiments were astounding. In applying electric currents to particular parts of the test animal's brain Ferrier was able to produce reflex acts and other physiological reactions. Another scientist who studied the functions of the nervous centres with the help the vivisectional method was William James, who pursued extensive experiments with frogs, severing their nerves at various points. In thus interrupting the flow of the nerve currents by cutting off the nerve centre from the rest of the organism, James was able to observe the locomotive reactions in the test animals, whose behaviouristic range was severely restricted and who were reduced, to use James's expression, to "an extremely complex machine" that "seems to contain no incalculable element" (I: 30).

Wharton disagreed with vivisectionist research and with inflicting pain on animals in the name of scientific research. Hermione Lee, for instance, notes Wharton's childhood preoccupation with animal and particularly canine welfare and comments on her involvement with the New York Society for the Prevention of Cruelty to Animals in the winter of 1905–6 (40, 206–7). Unsurprisingly, therefore, certain parts of *Ethan Frome* read like an antivivisectionist protest. We can see this, for instance, in the description of Ethan's paralysis when he is confronted with Zeena's decision to send Mattie packing: "His wife's retort was like a knife-cut across the sinews and he felt suddenly weak and powerless" (78). While the power struggle between the couple assumes the characteristics of a Jamesian vivisectionist experiment, it is suggested that Ethan's future plans – all of which involve the pleasant vision of Mattie's permanent presence in the Frome household – are rigorously interrupted by Zeena's hostility, which "had mastered him" (79). Confronted with his wife's "evil energy" (79) Ethan is, not unlike James's and Ferrier's laboratory animals, reduced to a helpless and passive machine, who can no longer interfere actively in the decisions that crucially affect his emotional comfort.

However, the showdown between Zeena and Ethan is by no means the only scene where Wharton drew on the scientific context that produced the vivisectionist experiments of Ferrier and James. It is worth remembering that Ferrier's test series with electricity, although it reduced animals to passive machines, also facilitated precise under-standings of the brain and the nervous system's integrative operations. And this reliance on electric currents to provoke physiological reactions also had other, very significant, theoretical and practical consequences. The first concerned therapeutics. After physicians such as Ferrier had demonstrated the dynamic effects of electricity on the animal nervous organization, electricity and electric gadgets started to be regarded as beneficial for the stimulation of an over-wrought nervous system. At the turn of the century both medical practitioners and patent medi-cal companies advocated electricity's potential in therapeutic practices. For instance, as John Harvey Kellogg, founder of the Battle Creek Sanatorium, suggested in his 1907 *Home Book of Modern Medicine*:

> Electricity is applicable — and more usually with success — to nearly all the curable diseases to which the human system is sub-ject.... Excellent results may be expected from the use of electric-ity in diseases due to or associated with general debility of the vital functions, impairment of nutrition, such as dyspepsia, neuras-thenia, or "nerve-tire," nervous debility, anemia, hypochondriasis, hysteria ... and various light affectations. (701)

Since medical practitioners believed that nerves were in "a continual state of vibration" and that the human organism experienced a change in "their electrical conductivity" as painful (Humphris 81), the sug-gestion that electrical stimulation would reverse this "altered electrical resistance" (Humphris 81) and remove pain must have appeared very plausible. Consequently, medical textbooks recommended a range of electro-therapeutic treatments and devices: hydro-electric baths, "the vibrating chair" (Kellogg 1621), electric belts and the vibrator (Figure 5) – a machine first designed in the early 1880s by Joseph Mortimer Granville for "electric or vibratory massages" (Dugan 177).[8] In Whar-ton's text, among the many costly remedies that Zeena purchases on her regular health check expeditions to Springfield, is "an electric bat-tery of which she had never been able to learn the use" (42). Zeena, Wharton's text suggests, has succumbed to the turn-of-the-century

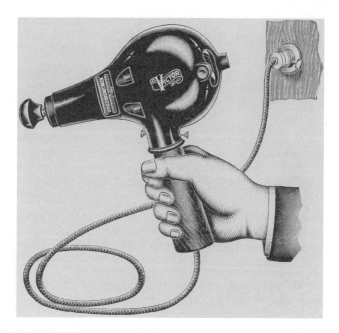

Figure 5 From William James Dugan, *Hand-Book of Electro-Therapeutics*, 1910, 179.

medical propaganda that advertised and promoted the restorative and therapeutic ability of electricity.[9]

In its depiction of the burgeoning relationship between Ethan and Mattie Silver, *Ethan Frome* similarly trades on the image of pulsating electric impulses. The physical and emotional vigour that "the sweetness of this communion" (22) produces is repeatedly described by evoking the image of electrical currents. Enjoying the intimacy of an evening alone (made possible by one of Zeena's therapeutic excursions), the lovers face each other across a table. But since they are too timid to touch or openly acknowledge their emotional and sexual interest in each other, Wharton's text indicates their mutual attraction with the help of an imagined electric stimulus. And it is Mattie's needlework that provides the base for the characters' brief moment of anastomosis:

> She sat silent, her hands clasped on her work, and it seemed to him that a warm current flowed toward him along the strip of stuff that still lay unrolled between them. Cautiously he slid his hand palm-downward along the table till his finger-tips touched the end of the stuff. A faint vibration of her lashes seemed to show that she was

aware of his gesture, and that it had sent a counter-current back to her; and she let her lands lie motionless on the end of the strip (63).[10]

In this emotionally charged scene, Wharton used the widespread belief in the invigorating force of electrical energy to translate her characters' psychological processes into clearly observable physiological sensations. And once again, as with Zeena's stifled complaints, Wharton, in this depiction, borrowed heavily from contemporary neuroscientific research and made use of its vocabulary. However, whereas an inappropriate form of nervous vibration creates Zeena's neurasthenic condition, the image of the electric current establishing a susceptible link between Mattie and Ethan references an unaddressed emotional rapport that receptively communicates any "alteration of mood" (64). Unsurprisingly, therefore, it is only when "the door of her room had closed" behind Mattie, that Ethan notices "that he had not even touched her hand" (65).

Quickened heartbeats and trembling lips: Emotions, willpower and Wharton's naturalist aesthetics

All this indicates how thoroughly Wharton relied on contemporary neurological and neuroscientific research in the writing of *Ethan Frome*. Its vocabulary gave her the opportunity to flesh out the novella's atmospheric setting. But Wharton's interest in neurophysiology and neuropathology also influenced the development of her literary aesthetics. I have already noted how decisively Wharton drew on the image of the lesioned test animal in characterizing her protagonist. Additionally, as will become apparent, the scientific context generated by Ferrier's and James's vivisectionist experiments gave her the opportunity to build up the novella's naturalist substratum.

Ferrier's experiments, especially his use of electricity in controlling test animals' movements and behaviour, had substantial epistemological consequences. As Laura Otis suggests, this particular type of vivisectionist research was so disconcerting for contemporaries because it implied that "electricity could replace a creature's will" (Otis 31). Ferrier, it was believed, was, with the help of electricity, able to control an animal's consciousness and willpower. This invited, on the one hand, speculations about the "physiology of emotions" (Wilson 356) and gave, on the other hand, impetus to understanding human consciousness, in its entirety, in predominantly mechanistic terms. Scientists asked the unsettling question: In what ways were human beings really comparable to machines? Thomas Huxley, for instance, was a fervent advocate of epiphenomenalism: the theory that consciousness is nothing but an epiphenomenon of the brain's or the nervous system's physiological processes. However, both inside and outside academic circles the question of how exactly physiological and psychological processes overlapped or disqualified each other in the human mind was fiercely debated. And predictably, deterministic readings of consciousness that presented human beings as reflexive machines were widely discussed in the turn-of-the-century print media. "Cells in the centre, and fibres running to and fro, constitute the spinal cord with its nerves", one columnist wrote in the *Dublin Review* in 1885, just to continue with the question: "But what have cells and fibres to do with thought, with love, with moral choice, with will?" (Sibbald 381)

Some neurologists answered this question by stressing the technological aspects in human behaviour. William James is a case in point. Although he clearly distinguished the mechanistic organization and reflexive behaviour of the central nervous system from the higher workings of the human mind that he saw as able to act spontaneously, independent from external influences and according to the particular psychical inclinations of consciousness, James, in his theory of emotions, seemed to have leaned more towards a materialist philosophy. Even though he did not deny a certain mental or spiritual component in the emotional experience, James, in thinking about "coarser emotions", emotions with a "strong organic reverberation" (II: 1065), defined them clearly as physiological upheavals or "reverberation" (II: 1066):

> Our natural way of thinking about these coarser emotions is that the mental perception of some fact excites the mental affection called the emotion, and that this latter state of mind gives rise to the bodily expression. My theory, on the contrary, is that *the bodily changes follow directly the perception of the exciting fact, and that our feeling of the same changes as they occur IS the emotion.* (II: 1065)

For James, physical sensations defined an emotional experience. If one subtracted them the emotional experience would become non-existent:

> What kind of an emotion of fear would be left if the feeling neither of quickened heart-beats nor of shallow breathing, neither of trembling lips nor of weakened limbs, neither of goose-flesh nor of visceral stirrings, were present, it is quite impossible for me to think. Can one fancy the state of rage and picture no ebullition in the chest, no flushing of the face, no dilation of the nostrils, no clenching of the teeth, no impulse to vigorous action, but in their stead limp muscles, calm breathing, and a placid face? (II: 1067–8)

Surprisingly, James, at a time when scientists had gained a thorough understanding of the internal workings of the human body and its nervous system, and who himself scrupulously dissected the brains of his test animals, preferred to read emotions exclusively from the surface of the body. Trembling muscles, a flushed face and shallow or calm breathing are all observable without the intervention of the microscope, the scalpel or the help of x-ray photography. But what must look like a scientific relapse indicates exactly the analytical insecurity, not just on James's but also on turn-of-the-century neuroscience's part, about how exactly neurophysiology and psychology could be brought together. A *Belgravia* correspondent summarized the predicament in 1881 as follows:

[T]he nature of the mental act — which is by some authors exclusively named the emotion — may be, and generally is, imperfectly understood by us; and the name is given rather to the obvious effects of the mind's action on the face and body, than to the mental action which is the cause of these visible effects. (Wilson 349)

Scientists remained baffled by the question of how exactly emotions were produced in the nervous system. Moreover, the consideration of emotions in neuroscientific research resolutely questioned theories that insisted on the purely mechanistic organization of the human nervous system.

Wharton, in spite of her apparent dislike for William James, seemed to have accepted this widespread trend towards examining emotional experiences in materialist terms on the body's surface. Rather than minutely describing her characters' emotional states, she focused predominantly on their physical reactions to external stimuli. The embarrassment Ethan feels when asking "for an advance of fifty dollars" is indicated only by the blood that "rushed to his thin skin" (50). Likewise Mattie's desperation about her uncertain future is referenced by "a tremor" that "crossed her face" (81) and readers can observe how Ethan's "tenderness" for Mattie "rushed to his lips" (90). Even the most violent emotional reactions remain connected to physical processes: "For a moment such a flame of hate rose in him that it ran down his arm and clenched his fist against her" (79). Thus Wharton, abhorrent of scientific speculation and analytical conjecture, registered, not unlike William James, her characters' emotional experiences solely by jotting down their external, physiological markers.

But although Wharton, in her quasi-clinical approach to representing emotionality seemed to have followed James, her decision to focus on the external expressions of characters' emotions, simultaneously criticized his reliance on vivisectionist experiments. Very likely, Wharton's objection to minutely describing characters' interior processes suggests a synchronized reluctance to dissect and expose the emotional currents of her literary protagonists in a way reminiscent of the baring and manipulating of the brains and nervous systems of James's and Ferrier's test animals. Furthermore, we can also relate Wharton's apparent dislike for prying into her characters' psychological landscapes to her rigorous antimodernist stance. In objecting to giving priority to the exploration of an individual's consciousness afforded by such contemporary novelists as Joseph Conrad or Ford Madox Ford, Wharton once again showed her dislike for modernist concerns. Indeed, her aesthetic allegiance seemed to have laid elsewhere.

As we have seen, turn-of-the-century neuroscientific theories emphasized the mechanistic elements in human physiology. *Ethan Frome*, while trading on these particular scientific contexts, also demonstrates a certain apprehension that human beings could be preset like machines. And such apparent misgivings bring Wharton's novella into contact with prominent late nineteenth-century concepts of biological determinism – pessimistic social theories that expressed scepticism about a subject's free will. Writing in a popular newspaper in 1885, a scientific correspondent summarized this determinist argument as follows:

But, in truth, the more closely we scrutinize our mental powers, and note the laws they follow, the more we are struck with the narrow limits within which our own action is restricted. To a large extent we are passive, and rather suffer our thoughts than think them. We may even more strictly be said to suffer than to do a large proportion of our own actions. (Sibbald 388)

"Much of our life passes before us like a panorama," the correspondent then pessimistically concluded, "in which we are rather the most interested of spectators than the actors" (388).

In Wharton's text Ethan's blatant and enervating passivity, his emotional and psychological restrictions and the novella's degeneration subtext, referenced by Ethan's father going "soft in the brain" (9), all indicate the characters' biological conditioning, their lack of determination and the individual's social paralysis in relation to powerful environmental forces. Unsurprisingly, Ethan feels "like a prisoner for life"

(89). He also feels that his dead ancestors in the graveyard mock "his desire for change and freedom". Indeed, their gloomy prediction, "[w]e never got away — how should you?" (34), amply illustrates the lack of alternatives in the lives of Wharton's characters. For instance, Mattie's half-hearted attempt to support herself by white-collar work inevitably ruins her health (40). And Ethan's "[c]onfused motions of rebellion" (87), which encourage him to consider his options, inescapably bring "home to him the relentless conditions of his lot" (88). Social roles and personal prospects are inevitably conditioned by biological and economic constraints in *Ethan Frome*. And Wharton's novel underlines this lack of social and economic progress further by commenting on her characters' locomotive stasis. Train lines are interrupted, movements are slow and mobility is generally obstructed in *Ethan Frome*. Ironically, only the suicide scene describes clear forward motion. However, the image of this apparent forward motion only reinforces the idea of predetermination, suggesting that Ethan and Mattie "follow the tracks" (112) of other sleds in order to crash into the elm tree. Like the linear sled tracks in the snow their future is predetermined, its course irreversible. In her most Calvinist narrative, Wharton therefore embraced those aspects of neuroscientific and neurological research that stressed the mechanistic elements in the human organization and that spurred such cultural proposals as the bleak prospect of evolutionary predetermination. Those selective components of the scientific debate corresponded well with her naturalist stance. But while Wharton unmistakably embraced neuroscientific and neurological paradigms, her novella also hints at the conceptual and analytical limits of modern sciences. *Ethan Frome*, as we shall see, forcefully questions science's ability to provide conclusive readings of human interiors.

Challenging the scientist in *Ethan Frome*: Wharton and modernism

Wharton's adaptation of neurological contexts, which correspond so neatly with the novella's naturalist orientation, suggests not only placid acceptance but also critical analysis. In spite of appropriating specific elements of contemporary neuroscientific research, Wharton, in *Ethan Frome*, also raised doubt about the future course of modern science. And paradoxically, with this decision to destabilize scientific hegemonies also came the resolution to critically interrogate her naturalist stance.

The most significant scientific agent in Wharton's novella is clearly the engineer narrator, who reads volumes of "popular science...on some recent discoveries in bio-chemistry" (11) and who investigates the Fromes with an indefatigable analytical diligence. He introduces himself as a scientifically detached observer with an apparent interest in "facts" (5, 7). He is also willing to regard Mrs. Ned Hale as another objective observer of the Fromes and comments favourably on her "finer sensibility and a little more education" that "had put just enough distance between herself and her neighbours to enable her to judge them with detachment" and to "co-ordinate the facts" (7). In emphasizing those activities, "judging with detachment" and "coordinating facts" in his own search for epistemological certainty, Wharton's engineer manages to give his curiosity about Ethan Frome the appearance of a scientific study with obvious anthropological and psychological components. The narrator clearly suspects that a human drama is responsible for turning Ethan into this "ruin of a man" (3) and for driving Ethan "too deeply into himself for any casual impulse to draw him back" (12). Both this mystery and Ethan's taciturn "silence" (10) motivate the scientific observer to investigate the details of this strange case.

Interestingly, in this attempt to analyze the Fromes Wharton's narrator sets up a clear distinction between factual analysis and storytelling, between facts and fiction. His narration opens with the words: "I had the story, bit by bit, from various people and, as generally happens in such cases, each time it was a different story" (3). Because of the different, conflicting interpretations of the Frome case, Wharton's engineer narrator is, right from the start, confronted with analytical ambiguity and uncertainty. Narratives and stories allow for interpretation and disagreement. The narrator tries to overcome this analytical uncertainty by collecting information and details about the Fromes. However, this attempt to recover the missing facts, his attempt to complete the case with certitude, is repeatedly frustrated. In trying to piece the story together our observer must finally accept that the "deeper meaning" of the Frome story "was in the gaps" (5). Like Zeena's "pale opaque eyes which revealed nothing and reflected nothing" (115), the Frome case remains inscrutable. The narrator's access to crucial details is denied and a conclusive interpretation becomes impossible.

In spite of this, Wharton's scientific observer nonetheless claims, self-righteously, after his encounter with the entirety of the Frome household, that "I found the clue to Ethan Frome, and began to put together this vision of his story" (17). Firmly trusting in his own analytical skills, the narrator then develops his version of the Frome case – the core narrative that occupies the biggest fraction of the novella. And it seems as if Wharton's narrator contrasts his discovery, his "clue", with all the other conflicting stories about the Fromes. Paradoxically, though, his words instantly question the validity of the following analysis. Although he tries to introduce his perspective as scientifically detached, objective and conclusive, Wharton's engineer simultaneously indicates that his reconstruction of the case is his "vision of the story". 'Vision' is hardly a scientifically exact term. It is a surprising semantic choice and one that is oddly reminiscent of William James, whose alleged scientific imprecision and analytical haziness was such a thorn in Wharton's side. It seems therefore as if Wharton's scientific narrator assumes the traits of a Jamesian psychologist, leaving the ground of exact science to adopt a perspective that Wharton would have dismissed as speculative. What is first presented as scientific certainty, based on clues and facts, turns out to be psychological speculation. In fact, the reader is offered yet another creative and subjective interpretation of the Fromes. In providing his "vision", Wharton's narrator therefore moves away from facts and factual investigation and embraces exactly the kind of interpretative ambiguity that he detected in the villagers' gossip and that provoked his investigation in the first place.

While the engineer's attempt to distinguish between a scientific and a fictitious interpretation of the case is thus problematized, readers are invited to question the accuracy of the core narrative and the narrator's reliability. His version of the Frome case is only one way of interpreting events – and one that readers might want to challenge. For instance, while the narrator is fundamentally sympathetic to Ethan's plight, he appears equally antagonistic in his description of Zeena and her "troubles". His androcentric perspective romanticizes Ethan's emotional paralysis and his lack of determination whereas Zeena's side of the story remains crucially underdeveloped. Clearly, the presented account raises as many questions as it answers. It is fragmentary, incomplete and contributes just another strand to the developing Frome myth. Hence in spite of many neuroscientific and neurological references, the promised scientific case study, based on facts and rational analysis, is not delivered.

But if the narrator's attempt at interpreting the Fromes cannot be regarded as a reliable recovery of 'missing facts', Wharton also reinstates creative interpretation as an analytical option. Readers should therefore feel invited to reflect further on the core narrative's "deeper meaning". Given Wharton's interest in human interiors, we can, for instance, read the narrative's ambiguousness metaphorically. Wharton was, as we have seen, sceptical about the analytical accomplishments of psychology as practised by William James. It is therefore not surprising that *Ethan*

Frome, in inviting its reader to question the accuracy of the core narrative, points subversively to limitations in the ambitious project of mapping psychological cores. Wharton herself preferred to represent characters' emotional experiences exclusively through exterior referents on their physiological shells. Deeper geographies of the human mind are left uncovered.[11]

More importantly, though, Wharton, in voicing reservations about the interpretative accomplishments of her scientific narrator, also destabilized the perspective of scientific objectivism that her novella seems to sanction. In *Ethan Frome* confidence in scientific rationalism gives way to interpretative uncertainty. And so, it is in the attempt to depict "the derelict mountain villages of New England", those "grim places, morally and physically" that were characterized by "insanity, incest and slow mental and moral starvation" (*A Backward Glance* 293–4), that Wharton, within the framework of her naturalist stance, drew closest to what she herself criticized most ferociously in the literary aesthetics of modernism: a pronounced nervousness and suspicion that the intellectual bedrock of empirical rationality is volatile and analytically insufficient in its confrontation with the neurological pathologies of modernity. On the textual surface of *Ethan Frome* Wharton's naturalist philosophy therefore mediates modernist preoccupations, producing a productive tension between competing literary philosophies that resembles, in many ways, the physiological stress produced by Zeena's agitated nervous system.

Notes

1. In her 2007 biography Hermione Lee similarly concludes that "the evidence for the Weir Mitchell cure is inconclusive" (79). That Silas Weir Mitchell's rest cure had, at times, counterproductive consequences for the patient's cure can be seen in Charlotte Perkins Gilman's *The Yellow Wallpaper* (1892).
2. A commonplace book is a form of scrapbook used specifically for educational purposes, defined by the *OED* as "a book in which 'commonplaces' or passages important for reference were collected, usually under general heads; hence, a book in which one records passages or matters to be especially remembered or referred to, with or without arrangement" ('Commonplace-book', *Oxford English Dictionary*).
3. In the *Principles of Psychology* William James famously argued that: "[c]onsciousness, then, does not appear to itself chopped up in bits.... It is nothing jointed; it flows. A 'river' or a 'stream' are the metaphors by which it is most naturally described" (I: 233).
4. James acknowledged the loosely structured organization of *The Principles of Psychology* in its preface: "The reader will in vain seek for any closed system in the book. It is mainly a mass of descriptive details, running out into

queries which only a metaphysics alive to the weight of her task can hope successfully to deal with." (I: 7)

5. Wharton was, throughout her career, interested in exploring the complex relations between the composition of interiors and their external referents, whether in psychological or indeed architectural terms. Accordingly, she lamented, in her first book *The Decoration of Houses* (co-authored with Ogden Codman), the disruption of "the natural connection between the outside of the modern house and its interior" (1). And in *Ethan Frome* the narrator comments explicitly on the contrast between Ethan's "outer situation and his inner needs" (12).

6. Interestingly, though, the expression 'mind' is, in one of the novella's many recourses to sentimentalism, used to describe Ethan's frustration about the impermeability (and irrationality) of Mattie's thoughts: "The motions of her mind were as incalculable as the flit of a bird in the branches" (32).

7. Elaine Showalter explains that the word 'receiver' is a "term from physics to describe the bell-jar used to create a vacuum" (*Ethan Frome* 121).

8. Many turn-of-the-century medical textbooks actually insisted on the difference between mechanical and electric vibration:

> Before entering upon this subject, however, it is necessary to insist upon one point, *viz.*, that mechanical vibration does not belong to the phenomena of nervous electricity. ... In short, electricity may take the place of nerve energy, even temporarily producing this energy, but it exerts no modifying influence on the nervous system, or more particularly, on the nerve centres. (Snow 10–11)

The cultural history of the vibrator as a therapeutic device, especially its use in so-called female pathologies, is discussed in detail in Maines, *passim*.

9. Interestingly, Wharton herself referred to the invigorating and revitalising effect of electricity in her letters when commenting on the construction of The Mount, her residence in Lenox, in 1902: "Lenox has had the usual tonic effect on me, & I feel like a new edition, revised & corrected, in Berkeley's best type. It is great fun out at the place, now too — everything is pushing up new shoots — not only cabbages & strawberries, but electric light & plumbing. I really think we shall be installed — after a fashion — by Sept 1st." (7 June 1902, *Letters* 66)

10. Wharton, in her journal, used a similar image to comment on her relationship with her lover, Morton Fullerton: "I felt for the first time that indescribable current of communication flowing between myself and someone else — felt it, I mean, uninterruptedly, securely, so that it penetrated every sense and every thought" (qtd. in Benstock 180).

11. It should not be forgotten that Wharton continued to demonstrate a keen interest in metaphysical matters. Commenting on the "celebrated reply" to the question, 'Do you believe in ghosts?' – "No, I don't believe in ghosts, but I'm afraid of them", she wrote: "To 'believe', in that sense, is a conscious act of the intellect, and it is in the warm darkness of the prenatal fluid far below our conscious reason that the faculty dwells with which we apprehend the ghosts we may not be endowed with the gift of seeing" (*Ghost Stories* 1). Wharton thus indicated her belief that certain parts of human interior geographies could not be explored with science's efforts.

Works cited

Benstock, Shari. *Edith Wharton: No Gifts from Chance*. London: Hamish Hamilton, 1994.

Cahir, Linda Costanzo. 'Edith Wharton's *Ethan Frome* and Joseph Conrad's *Heart of Darkness.' Edith Wharton Review* 19.1 (Spring 2003): 20–3.

'Commonplace-book.' *Oxford English Dictionary*. 2nd ed. Oxford: Clarendon, 1989.

Dugan, William James. *Hand-Book of Electro-Therapeutics*. Philadelphia: F. A. David, 1910.

Ferrier, David. *The Functions of the Brain*. London: Smith, Elder and Co, 1876.

Fryer, Judith. *Felicitous Space: The Imaginary Structures of Edith Wharton and Willa Cather*. Chapel Hill: North Carolina UP, 1986.

Granville, Joseph Mortimer. *Nerves and Nerve-Troubles*. London: Allen & Co, 1884.

Humphris, Francis Howard. *Electro-Therapeutics for Practitioners*. London: Edward Arnold, 1913.

Hutchinson, Stuart. 'Unpacking Edith Wharton: *Ethan Frome* and *Summer.' The Cambridge Quarterly* 27.3 (1998): 219–32.

James, William. *The Principles of Psychology*. 3 Vols. Cambridge, Mass.: Harvard UP, 1981.

Kellogg, J.H. *The Home Book of Modern Medicine: A Family Guide in Health and Disease*. London: C.D. Cazenove, 1907.

Lee, Hermione. *Edith Wharton*. London: Chatto & Windus, 2007.

Lewis, R.W.B. *Edith Wharton: A Biography*. London: Constable, 1975.

Maines, Rachel P. *The Technology of Orgasm: "Hysteria," the Vibrator, and Women's Sexual Satisfaction*. Baltimore: Johns Hopkins UP, 1999.

Otis, Laura. 'Howled Out of the Country: Wilkie Collins and H. G. Wells Retry David Ferrier.' *Neurology and Literature, 1860–1920*. Ed. Anne Stiles. Basingstoke: Palgrave, 2007. 27–51.

Peel, Robin. *Apart from Modernism: Edith Wharton, Politics, and Fiction before World War I*. Madison: Fairleigh Dickinson Press and Associated University Presses, 2005.

Ramón y Cajal, Santiago. *Histology of the Nervous System of Man and Vertebrates*. Trans. Neely Swanson and Larry W. Swanson. 2 Vols. New York: Oxford UP, 1995.

Sibbald, Andrew. 'The Brain and the Mind.' *Dublin Review* 3.13.2 (1885): 381–92.

Singley, Carol J. *Edith Wharton: Matters of Mind and Spirit*. Cambridge: Cambridge UP, 1995.

Snow, Mary Lydia Hastings Arnold. *Mechanical Vibration: Its Physiological Effects: Its Therapeutic Use: Its Practical Application*. London: Henry Kimpton, 1906.

Wharton, Edith. *A Backward Glance*. London: Constable, 1972.

——. *Ethan Frome*. Ed. Elaine Showalter. Oxford: Oxford UP, 1996.

——. *The Ghost Stories of Edith Wharton*. London: Virago, 1996.

——. *The Letters of Edith Wharton*. Eds. R.W.B. Lewis and Nancy Lewis. New York: Collier, 1988.

Wharton, Edith, and Ogden Codman. *The Decoration of Houses*. New York: Norton, 1998.

Wilson, Andrew. 'The Mind's Mirror.' *Belgravia: A London Magazine* 45.179 (1881): 346–66.

10
Sounds of Silence: Aphasiology and the Subject of Modernity

Laura Salisbury

On the face of it, there isn't much to be made of a broken hammer. No longer functioning as a tool that can be used to mould and shape the world, the broken hammer sits there mutely, with a seemingly obdurate, obstinate, materiality, refusing to become a smooth extension of the hand working with intention. A hammer that cannot be used as such is, in some senses, no longer a tool at all; it is just one particular configuration of metal and wood set apart from human projects. For Martin Heidegger in *Being and Time* (1927), however, a broken tool was not just one material object amongst others; for in deficit, a broken hammer could articulate something of the world of human involvements in which it took its place as an instrument. Indeed, the damaged thing and our relations with it suddenly seemed 'obtrusive', as the tool's 'unreadiness-to-hand' roughened up the world of complex spatial and temporal involvements that was previously so smoothly clear for human attention, forcing it against the grain of habitual perception. For Heidegger, then, the gap left by the broken hammer revealed something that was normally so transparent that it hid its character; it demonstrated that the human subject was necessarily and immanently connected to a world that was determined by networks of intentions and involvements: "Our circumspection comes up against emptiness and now sees for the first time *what* the missing article was ready-to-hand *with*, and *what* it was ready-to-hand *for*. The environment announces itself afresh" (105).

That what was refractory could also be revelatory became a trope that emerged to dominate epistemologies of modernity. Of course, negation had classically functioned as an important philosophical method; but modernity's obsession with sensations of newness, an awareness of a 'just now' that compulsively sought either to transcend a seemingly less developed past or began, with a new and distinctive fervour, to bemoan the loss of any smooth connection with a previous 'golden age', produced ways of understanding the world that were structurally traced out by the contours of loss and lack, newly estranged from what had seemed like previously transparent narratives of self-legibility. But if this is so, it is hardly surprising that medicine, which has always been concerned with loss, injury and disease, offers one of the clearest artic- ulations of modernity's tensely ambivalent concern with negation and the revealing gaps of deficit. For medicine necessarily attends to experi- ences of body and mind that are somehow roughened up or impeded, gleaning knowledge of the workings of soma and of psyche from those natural experiments arising from the effects of disease or wounding. Although medical history is replete with positivist and interventionist discourses that emerged from both laboratory and clinic, in the accounts of the material production of language within the brain that emerged from modernity it was the observation and interpretation of deficit that remained central. This chapter will argue that the call to articulate an account of healthy linguistic function from a newly traceable contin- uum with the pathological replaced the idea that it was possible to intuit a self-present speaking subject whose language was immanent inside it and that functioned as a smooth transcription of a coherent and ratio- nal intentionality. From the discourse of aphasiology there emerged, instead, an account of language and a speaking subject within which meaning did not inhere, it could not simply be found; it needed instead to be forged and made.

Aphasia can be defined, somewhat imperfectly, as the "loss of speech, partial or total, or loss of power to understand written or spoken lan- guage, as a result of disorder of the cerebral speech centres" (*OED* 546), and it holds a rather singular position within the discourse of neurol- ogy. For although humans are not the only animals to utter sounds of communication, it has nevertheless been clear, since ancient times, that people are the only animals possessed of the capacity for propositional speech; indeed, it is this speech that has often been taken to define the locus of the human. But it has also persistently been clear that it is pos- sible for a subject to be denuded of this capacity, due to the natural experiments of disease or wounding. If the human is the only animal that talks, however, there are no other species upon which experimen- tal productions of language deficit can effectively be performed (at least not in the period with which this volume is concerned): no frogs to be galvanized into speechlessness, no cats, pigeons or monkeys, who could be subjected to an experimental wounding that would reveal very much about diminished linguistic capacity. Even within a period so dominated by technological advances and empirical experimentation, studies of the functioning aphasic brain were necessarily confined to live humans in whom diminished capacity could not ethically be precipitated, even if the aphasic patient's speech could be studied and then the loss of function mapped on to damaged areas of the cortex post mortem.

Language was, and of course remains, a precipitate of the brain that does not persist after death. Although a technique of hardening the brain in alcohol had been developed by Reil in 1809 to allow its adequate dissection (Clarke and Jacyna 99), unlike those nerve fibres that control movement, there was no possibility of pumping impulses through dead brain matter that could reveal anything of conceptual and linguistic functioning. Indeed, it was not until the late twentieth century that functional magnetic resonance imaging and computerized axial tomography offered any hope of dissolving the occlusions of the cranium to enable the observation of healthy linguistic functioning within the brain. In this period of modernity, then, aphasiology offered insights into language's production in the brain that were necessarily dependent upon negatively interpreting the shapes and processes of healthy function from the manifestations of disease, from listening to and interpreting the noisy utterances of the aphasic patient. As with Heidegger's illuminating broken hammer, it was only the shock of aphasic deficit caused by disease or wounding, the machine suddenly revealing its cogs as it began to break down, that sufficiently displaced habitual models of perception to reveal normative linguistic brain function in all its unexpected strangeness. It was through deficit and damage, according to the troubling uncertainty and clamour of a language rendered suddenly strangely opaque, that the speaking and indeed the thinking subject of a newly neurological modernity appeared.

Locating language

The brain had long been posited as the seat of cognition, and a number of eighteenth-century scientists had proposed determinedly physiological models of mind. Despite strong resistance from writers who affirmed idealist and transcendental conceptions of thought, with various post-Cartesian dualisms recoiling at newly corporealized accounts of the mind, the diverse physiological 'Romantic psychologies' of Erasmus Darwin, Gall, Cabanis and Bell, as Alan Richardson has clearly shown, all located the brain as the 'cerebral organ' (1–38, 66–92). But linguistic function continued to remain a fly in the materialist ointment, for within broader philosophical discourses there persisted the notion that language, as the house of reason, was that which distinguished human from animal, and was thus decisively linked to the immaterial part of human nature. As in ancient philosophy, because it was the only animal that spoke with seeming intention, the human was distinctive: it remained, singularly, *zoon logon ekon*.[1] Descartes, as perhaps the most discursively powerful example, asserted that one might be able to construct a machine that could process food and function anatomically like an animal, and which might even be able to utter words, but it would not be able to arrange these words responsively and in rational, propositional terms, just as an animal could not. In 1649 he opined:

> it has never been observed that any animal has attained the perfection of using real speech.... Such speech is the only certain sign of thought hidden in a body. All human beings use it, however stupid or insane they may be, even though they may have no tongue or organs of voice; but no animals do. Consequently this can be taken as the real specific difference between humans and animals. (165)

For Descartes, as for many others, the origin of the propositional speech of the reasoning human persisted as a site in which *deus* persistently found itself being inserted back into *machina*.

Despite complex debates on the relationship between 'natural language' and the 'arbitrary signs' of human speech, despite controversies as to whether language had been divinely imposed upon the animal and material part of the human or somehow arose from it,[2] L. S. Jacyna reveals in his illuminating history of aphasiology that

> a putative continuity between language, reason and humanity in the eighteenth and early nineteenth centuries persistently ensured that language served as a boundary category which reinforced a number of crucial hierarchies. Man's endowment with language distinguished him from the animal. Language indicated the duality of human nature – appertaining to a superior, spiritual aspect. The body was accordingly stigmatized as 'the lesser part' of the human constitution. (62)

Although it is important to register revisionist histories of the relationship between brain function and language that stress burgeoning physiological accounts – for the eighteenth and early nineteenth century was precisely the period when the theorizations of mental faculties as parts of an unextended, and therefore potentially transcendent soul, were losing their discursive force – it is also vital to note that the emergence of aphasiology in the 1860s as a discourse of modernity persistently figured itself as a materialist understanding of language that could be opposed to the immateriality of previous conceptions. As the neurologist Henry Head asserted in his account of the history of aphasia in 1926:

> [t]hroughout the eighteenth century the doctrine of the relation between mind and matter became increasingly metaphysical; soul and body were thought to be coexistent but independent factors in the life of man.... [T]he brain was looked upon as a single organ from which flowed vital energy under the influence of the will into all parts of the body. (*Aphasia* 3)

So despite registering the continuities with the complex neurology of the pre-nineteenth century, Head determined that aphasiology's theories of propositional language function, born in the 1860s, represented a scientific overcoming of an earlier account that had little material voice.

The year 1861 thus appears as something like a turning point in the history of language. Although earlier writers had posited the possibility of an organ of speech, the 1860s was the decade when the debate all but ceased in medical discourse as to whether language was produced and represented within the materiality of the brain; discussion focused instead upon precisely how this representation was to be theorized and modelled. The word aphasia was first used in 1864, following Paul Broca's (1824–80) famous presentation of a case of 'aphemia' in a patient called Leborgne in 1861.[3] Broca offered his case as part of a wider debate at the Societé d'anthropologie in the Paris of the Second Empire (1852–70) on the question of 'Big Heads', and whether there was a direct relation between brain size and intelligence – a point that, if proven, would ablate the putative boundary between spirit and matter. The debate soon centred on the question of language, which, as the highest of human faculties, could offer definitive proof of the localization of specific functions within the brain. Broca's belief in the principle of cerebral localizsation of faculties was demonstrated by the presentation of Leborgne's case. Paralyzed in the right leg and almost mute for 21 years, Leborgne had been able only to produce the utterance 'tan', alongside a few swear words and oaths. His brain, at autopsy, was discovered to have a lesion in the third frontal convolution of the left hemisphere caused by a cyst. This area was put forward as the localized centre for the 'faculty of articulated language' and Leborgne's brain presented for inspection to the Society. Following Broca and the fierce debates that succeeded this presentation, speech and language were both firmly assimilated into models of organic functioning, with empirical studies of the speech pathologies of brain-damaged patients, whose cerebra would than be analyzed at autopsy, disseminated through a well-developed international medical press. Within a few years of Broca's initial presentation of the 'Tan' case, the theory of cerebral localizations had been cemented within the power of scientific orthodoxy (Young 134), and as linguistic control and ability became discursively localized in particular areas of the brain, a new conception of speech and writing emerged which disarticulated the smooth and necessary connection between language, reason and the immaterial or spiritual parts of the human.

Noisy machines

One of the most suggestive readings of these new neurological conceptions of language and their relationship to the cultural discourse of modernity comes from Friedrich Kittler, who integrates both late nineteenth- and early twentieth-century aphasiology and modernist artworks into a cultural history of information. Kittler claims that the new writing and recording technologies of modernity fundamentally restructured general conceptions of language and, in turn, determined the ways in which language could be used to process information in the production of ideas and cultural artefacts. What Kittler calls the 'discourse network' of gramophone, film and typewriter of 1900 – that is, the "network of technologies and institutions that allow a given culture to select, store and process relevant data" (*Discourse* 369) – dictated both how this data could be expressed formally, and how ideas, as the processing of information, came into being.

Kittler's somewhat partial argument is that language in the late eighteenth and early nineteenth centuries was deemed, in most formulations, to function as a transparent representation of the transcendent inwardness of the soul. For Kittler, the acquisition of literacy in the 1800 discourse network underwrote a conception of language anchored to a repetition of the supposed plenitude of the mother's voice that required the perception of a natural, motivated link between written word and sound. In this period, new methods of creating literate subjects had replaced an emphasis on rote learning, imposing instead the necessity of understanding an immanent meaning, to which the reading and writing subject could gain access by attuning subjectivity to an internal voice that spoke in harmony with a natural language that pre-existed it. The writing hand, in turn, formed a smooth connection between linguistic production and the expression of this inner tran-scendent meaning, this language of nature. But as handwriting ceded dominance to new technologies such as the typewriter (first sold com-mercially in 1870) in the 1900 discourse network, "writing was no longer the handwritten, continuous transition from nature to culture. It became selection from a countable, spatialized supply" (*Discourse* 194). Unlike handwriting, the new technologies of the phonograph (1857) and film camera (1888) also recorded relatively indiscriminately what was within their range. Without having to submit information through the bottleneck of alphabetization, they blurred the distinc-tion between what appeared as random data and meaningful sequence. Because the proliferation of both intended and unintended signifiers clouded the transparency of the text as a clear window of represen-tation, the materiality of the medium and of signification itself came to the fore. The recording of both meaning and noise, sense and nonsense, thus restructured the idea of reading in modernity away from any inner apprehension of a transcendental signified into a pro-cess dependent upon decoding, upon interpreting the meaning of difference.

Kittler notes that where new technologies reconfigured language as information, discourses of brain localization emerging from aphasiology also found themselves suggestively mirroring the mechanical workings of the phonograph (*Gramophone* 38), or the typewriter. Carl Wernicke (1848–1905) offered perhaps the most influential account of localized brain function in the period, producing models and diagrams that reconstructed language, as an agent of consciousness, as a process of mechanical communication of sensory-motor units along 'association fibres'; he also compared damaged brains to malfunctioning telegraphs: "If certain letters are missing in the apparatus, specific errors would be consistently repeated in the message" ('Motor Speech' 152–3). Deter-mining that aphasia resulted from "a disruption of the psychic reflex arcs used in normal speech processes" and a disturbance of the pathways between sensory memory images and motor movement images that cre-ated associations, the brain damage traced out by Wernicke's classical model, and the insights into function it threw into relief, offered a way of conceiving of what was persistently to be referred to in the 1880s as the breakdown of a decidedly modern human machine (see Lichtheim 433–84).

Linking scientific discourses, new writing technologies and the linguistic turn of twentieth-century continental philosophy, Kittler suggests that the discourse network of 1900 reduced the wholeness of language to a knot of nervous sensory-motor fibres in its production as speech, and a scattered distribution of differential marks and traces in its appearance as writing. And only because aphasiology, following the forms of new writing and reading technologies, systematized and externalized language by severing the immutable connection between utterance and intention, did it then make "terminological sense", according to Kittler, "for Saussure, in founding a new linguistics, to decompose the linguistic sign into the concept (signified) and the acoustic-sensory image (signifier), or for Freud...similarly to divide 'thing representation' from 'word representation'" (*Discourse* 216). Although Kittler does not mention it, it seems significant that Broca needed to imply, through spelling, that Leborgne's 'speech automatism' 'tan' was a nonsense word. Of course, Leborgne's word could have been read differently, perhaps as a recognizable (although misused) signifier: 'tan' could have been 'tant', 'temps', 'ton', 'thoan', 'thon' (dependent upon regional accent).[4] In 1861 the aphasic speech automatism had to be a nonsense word because it was clearly unpropositional and thus needed to be divorced from intentional control and stable referentiality; but after Broca and after Saussure, recognizable speech automatisms could be invited back into the fold of language because *all* relationships between signifiers and signifieds were reconceived as arbitrary and necessarily conventional. Once the systematic and functional analysis of language had finally definitively loosened the signifier from the signified, describing the bond between them to be both arbitrary and relational in Saussure's famous terms, meaning necessarily proliferated and active interpretation assumed the place of apprehension as the privileged mode of reading within modernity.

Seeing sense

Although Kittler emphasizes the production of both signal and noise from those new writing machines that served as models for brain function, it is important to note that there was not much noise in classical aphasiology. Kittler is intrigued by the nonsense sounds emerging from the psychophyscial experiments of a scientist such as Gustav Fechner, as he tested the capacity of the mind to remember words alongside non-signifying syllables, but there is little evidence that classical aphasiology between the 1860s and the 1910s used much experimental methodology. Both Broca and Wernicke assessed patients through observation rather than testing, with patients' brains analyzed at autopsy and the locations of lesions or malformations noted in relation to deficits of function. Indeed, despite Kittler's insistence on the production of aphasic noise, the voice of the aphasic patient and the racket of aurality were almost completely written out of the clinical accounts of classical aphasiology, to be replaced with silent, dead anatomy and spatialized models that, with a primarily scopic drive, turned the aphasic brain into a predictable mechanism.

The desire to reproduce diagrams, quasi-algebraic formulae and maps of localized function within the cortex, asserted a positivist belief in a mathematical order that underlay the flux of the phenomenal, but in 1926 Head bemoaned the abstraction of these models after 1871, when "the rage for diagrams became a veritable mania" (*Aphasia* 103) that wrote over the complex materiality of the patient. For Wernicke, though, a unified, schematically coherent system of brain function remained a virtuous one that articulated its truth. Replete with diagrams of cortical functioning and quasi-mathematical formulae, Wernicke's work of 1874 asserted, with positivist clarity, that physiology's obsession with function could be matched by a clear perception of form; he determined that he had "illustrate[d] certain general principles based on current anatomical and physiological evidence, which may be applied to the lawful generation of spontaneous movement. All spontaneous motor action, including that of speech, is based on such principles" ('Aphasia' 69). Although even Wernicke's work included some case studies in which speech sounds and pathologies were reproduced (see 'Aphasia' 119–42), it was the spatialized diagram of localized brain function and a mechanically associating mind rather than temporal modulations of narrative that dominated. A view of brain function as that which could be possessed all at once, reproduced in a diagram, was laid over any long accounts of patients' speech, any reproductions of their noisy narratives that might display temporal twistings, expressions of subjective affect, or the complex recursions and progressions of brain damage as it began to heal itself. The silent rationality of the schema was thus won at the expense of the threatening complexity of the narrative torsions emerging from the material patient.

In *Suspensions of Perception*, Jonathan Crary has described how new modes of conceptualizing attention that emerged from experimental psychology, psychophysics and neurology in the latter half of the nineteenth century, fundamentally reconfigured dominant accounts of perception in modernity, blurring the Enlightenment belief in the stability of vision. He argues that eighteenth-century models of vision were determined by the metaphorical insistence of the camera obscura, "in which an ideal observer had the capacity to apprehend instantaneously the unedited contents of a visual field" (39–40). Objects of vision were represented as self-present to their observer, perceived in an instantaneous, atemporal and fundamentally spatialized process of apperception. But, in Crary's account, the rise of physiological optics in the first half of the nineteenth century displaced this model, suggesting instead that "a full grasp of self-identical reality was not possible and that human perception, conditioned by physical and psychological temporalities and processes, provided at most a provisional, shifting approximation of its objects" (4). Crary reads the 1850s–1870s as the moment when the vagaries of attention came to dominate accounts of mental processes, with perception and, by implication, even mentation itself, becoming an activity of selection and exclusion. Attention rendered the phenomena of the world perceivable by excluding what was on the fringes of its concern; physiology and psychology thus reconfigured the subject of modernity as one that was unable to achieve a self-presence with the phenomenal world, but was constantly shifting and sifting sense from a world with which it was engaged and simultaneously alienated in a temporal process of oscillating focus.

Wernicke's map-making appears, in this context, to offer a strong attempt to place aphasiology within rather older accounts of knowledge and vision, as the patient's body, their psychological processes and even the material functioning of the brain itself, were filtered out, occluded as noise within the system of what was a purer, more perfect visual field. Shaped against the contours of a more general epistemological uncertainty in which perceptual experience had lost its self-presence, its privileged apprehension of truth, Wernicke's aphasiology imagined a perfectly balanced equation of simple negation that refused to listen to the noise that seeped back into the system within the symptom of brain damage. As Head put it in 1915, for classical aphasiology "[e]ach patient with a speech defect of cerebral origin is stretched on the procrustean bed of some theoretical scheme; something is lopped away at one part, something is added at another, until the phenomena are said to correspond to some diagrammatic conception, which never has and never could have existed." ('Hughlings' 1) For Head, the new aphasiologist had to listen to the ways in which the subjective symptom introduced complexity into the system if a more accurate, flexible model of brain function was to be perceived.

Wresting human from machine

In his 1926 *Aphasia and Kindred Disorders of Speech*, Henry Head (1861–1940) reproduced an extract of a letter from a private soldier written in 1915:

> "My dear Sister, I now write to you that I hame getting the best of hearl and I am getting to walk to the garden to play a lot of game and I hame the all of." When the soldier was asked to read his letter back aloud to his doctor, he reproduced something quite different: "My dear Sister, I now write to you that I am getting the best of health and I getting to walk to the garden to play a lot of games and I have the whole of ... I've forgotten now what I was going to say." (77)

The soldier had been admitted to the London Hospital on 2 March 1915. In the last days of January he had been hit by a rifle bullet, which entered to the left of the inner canthus of the right eye and passed out just above the insertion of the left ear. Having mostly recovered from the obvious physical effects of his injury, he began to be treated in the London Hospital by Head. Head is probably best remembered for his work on the physiological basis of sensation and the inquiries into the sectioning and regeneration of cutaneous nerves (1903–7) with W.H.R. Rivers in which he became his own experimental subject; but during the war of 1914–18, he abandoned private practice to treat wounded soldiers at the London Hospital, and convalescing officers at the Empire Hospital in Westminster. Inspired by the writings of John Hughlings Jackson, Head began systematically to study and treat aphasic patients, and the two-volume work from 1926 in which this transcription of aphasic speech appears, is both a history of aphasiology from the position of post-war modernity and a record of the significant shifts in aphasiological discourse produced by the war.

If one of the narratives modernity tells about itself concerns an overcoming of the past and a movement towards the clear light of knowledge, neurology, alongside medicine in general, is a discourse that emerged from the wound of the First World War with its teleological trajectory of development surprisingly intact. In civilian practice, the majority of aphasic subjects presented to neurologists were those suffering from degenerative diseases, tumours or strokes. Manifesting the effects of unruly nature, of a disobedient materiality that seemed to divest the subject of its cultured rationality, its status as a *cogito*, aphasia was a disease at odds with a certain version of modernity, even if it could be ordered and controlled in the stasis of autopsy. But the War, as a factory production line of relatively discrete head wounds, created a new object of suffering – a new aphasic subject who had been produced by the shock of technological modernity, but who could be more easily encoded, whilst alive, into modernity's narratives of progression.

As Jacyna notes in his account of Head's work, advances in neurological knowledge were not simply products of the large quantity of subjects treated and the opportunities afforded for observing their often long recoveries (*Lost* 150–6); Head was more engaged by the singular *quality* of these other 'Men of 1914':

> In civilian practice many of those who suffer from aphasia are old, broken down in health and their general intellectual capacity is diminished.... But the war brought under our care young men who were struck down in the full pride of health. Many were extremely intelligent, willing and anxious to be examined thoroughly. As their wounds healed, they were encouraged and cheered by the obvious improvement in their condition. They were euphoric rather than depressed, and in every way contrasted profoundly with the state of the aphasic met within civilian practice. (*Aphasia* 146)

German aphasiologist Kurt Goldstein (1878–1965) concurred in *After-effects of Brain Injury in War* (1942), which detailed his treatment of soldiers injured in the First World War: "We are not dealing with progressive diseases, but with young men with long lives before them.... The patients themselves show more willingness to learn, and there is much more hope of improving their mental condition, than in the case of diseased individuals" (66). These were self-determined young men, often of officer class, whose previous intellectual and linguistic capacity was easily reconstructed according to class norms and military records.[5] They were also eager to place themselves within a regime of scientific testing and treatment, and their symptoms were more discrete and indeed more suitable to medical taxonomies than the complicated, progressively degenerating subjects found in civilian practice. Head noted that "with gun-shot wounds of the head the symptoms tend to clear up to a considerable extent, provided there are no secondary complications" because they occur due to penetration from without rather than the production of lesions or tumours from within (*Aphasia* 146). This enabled doctors to track and trace "the various steps by which defective functions are restored, whereas in civilian practice any change in the clinical manifestations is usually in an opposite direction" (*Aphasia*

146). Where a localizationist like Wernicke maintained a certain therapeutic nihilism, Head and Goldstein insisted that their military patients
assumed a legible position within a narrative of rigorous examination,
scientific testing and therapeutic progression.

The desire to place patients within a discourse of therapeutic heroism appeared persistently in the clinical accounts of both Goldstein
and Head. Goldstein asserted that society remained indebted to "young
men crippled for life by the impairment of man's most precious capacity, the power of mind, while serving their country and fellow-men"
(*Aftereffects* 223), whilst Head, as Jacyna notes, sent many letters to
his wife describing his patients as "dear young men" (*Lost* 152), possessed of physical and moral fortitude. Head had previously responded
to the requirement to produce accurate accounts of subjective symptoms by becoming his own experimental subject, but aphasic soldiers
now offered him instructive and admired objects of study who were
deemed to be educated, responsible and fundamentally rational subjects. Although defective of speech, they were not without language
completely, and their education, class position and masculinity made
them trusted witnesses of their own disability. Head's accounts of his
regime of testing and treatment indeed placed an unusual emphasis on
a new equality and reciprocal relationship between doctor and patient.
He describes how, during one of their sittings, Case No. 14 touched
the depressed part of his wound and then wrote 'forest branche'. Head
recounts:

> As I failed entirely to understand his meaning, he made me fetch
> from the cupboard in his room a book on forestry; he turned the
> pages until he came to a picture of a wood with undergrowth to
> which he pointed, saying 'branch,' at the same time touching his
> head. By this means he was able to make me understand his fear
> that, when walking through a coppice, the branches might injure his
> wound. He was at once satisfied when I told him that we would cover
> it with an aluminium shield. (*Aphasia* 239)

Within Head's work, the aphasic patient's subjective experience and
indeed the sounds of his disabled language were reproduced with a new
and distinctive centrality.

Head and Goldstein emphasized the persistence of personality, intellectual ability, interests and civilian skills, even a class identity, which
lay anterior to these brain-damaged soldiers' disabilities, as both doctors were keen to position them as coherent subjects who assumed their

own singular position within a rational and scientifically reproducible schema of treatment.[6] Within such a narrative, the patient became a subject whom the doctor was able, more or less successfully, to return to their previous and essential self through a regime of testing and treatment. These accounts of aphasia were infused with a humanism – an empathetic engagement with their subject rather than a simple account of symptom – and a certain faith in scientific positivism, then, as both doctor and patient were reconfigured as essentially coherent subjects with proscribed positions in a rational system. Clearly responding to the discourse of 'scientific management' and efficiency that was entering both the workplace and the clinic on the eve of the First World War, just as Taylorism itself had been influenced by psychophysics and the psychological laboratory (see Rabinbach 238–70), Goldstein established a diagnostic and therapeutic regime for the mass of brain-damaged soldiers by setting up the Institute for Research into the Consequences of Brain Injuries in Frankfurt in 1916. The Institute, which remained in operation until Hitler came to power in 1933, integrated the hospital for treatment, the psychological laboratory for individual testing and the school and workshop for retraining. Although neither Goldstein nor Head recommended that a brain-injured soldier should ever be returned to the Front, there was consistent emphasis on returning all but the most seriously disabled to work in order that they might utilize their surviving capacity. The twin goal of the treatment of aphasia thus became, in the work of these doctors, to restore the latent and pre-existing wholeness of the individual personality, and then to make him of use to society rather than a burden on its resources. The integration of the hospital, the laboratory and the workshop attested, then, to the increasingly continuous network that joined economic productivism, psychophysics, experimental psychology and neurology. Where psychotechnical testing, used obsessively during the early 1920s in Germany to assess the fatigue, trainability and performance of the body alongside occupational aptitudes and inclinations of the mind, was recommended as a rational method of returning healthy soldiers to the post-war economy,[7] the brain-damaged soldier, with training that took into account his latent proclivities, the effect of the disability and his reconfigured capacities, could similarly emerge as a productive subject of modernity.

The instrumentalism of the necessity of work should not, however, be separated from Goldstein's belief in its therapeutic benefits; for restoring a subject to economic and social usefulness was part of the process of reinstating the soldier's own rational self-legibility. But despite the emphasis on placing the aphasic subject back within the economic system, Goldstein worried that modern employment conditions were fundamentally unsuited to accommodating the unpredictability of brain-damaged workers: "big business and industry today are so highly organized, and demand such steady work in offices and on the assembly lines, that fluctuations and irregularities in work disrupt the whole routine" (*Aftereffects* 209–10). Goldstein's patients, it turned out, were more suited to less systematized, less mechanized and, indeed, less modern employment. The area of employment that offered the clearest potential was that of the "artisan, such as a locksmith, accustomed to work in a small business" (*Aftereffects* 213), which the workshop in the Institute replicated. At precisely the moment that theories of a romanticized return to pre-industrial work communities were emerging in Germany, such as Eugen Rosenstock's theory of 'Werkstattaussiedlung' (workshop-relocation), in which workers would be transferred from large urban plants to the countryside to form their own small groups and work in their own 'living space' (Rabinbach 282), the aphasic patient, as a victim of industrialized warfare, was recognized as a subject fundamentally unsuited to reintegration into an economy of industrialized modernity. Head similarly observed that his aphasic patients often displayed profound difficulties in telling the time or even predictably marking temporal duration; indeed, he was obsessed with noting the problems his patients had with understanding British summer time, newly instituted in 1916.[8] The aphasic patient clearly had an untenable position within a scientifically managed economy based upon stable units of temporal exchange.

The complex and excessive qualities of the manifestations of aphasia, even within the soldiers whose discrete wounds and educated self-presence seemed to make them the perfect exemplars of the pathology, thus persistently seeped through into Head and Goldstein's accounts. For however many models and diagrams were made of the placement of particular brain lesions, the aphasic symptom remained relatively unpredictable, a narrative of loss but also a story of the plastic *adaptive reaction* to the experience of wounding, which disavowed the possibility of a simple hydraulics of negation, in which a loss here would create the appearance of a predictable and immediately visible symptom there. Aphasia, as Head and Goldstein asserted, was a complex constellation of symptoms and psychical compensations that were only triggered by the organic injury which could be observed and mapped onto the brain. It is highly suggestive that despite their desire to preserve the integrity of the man suffering an organic injury from the vagaries of psychological disorder,[9] both Head and Goldstein were insistent that it made little sense to make firm distinctions between psychical and somatic processes in the analysis of compromised language and cognition caused by brain damage. Goldstein admitted that the whole dichotomy of organic/psychogenic pathology was at fault, stating that the "organic patient will react to his condition with functional symptoms also. Furthermore, the functional symptoms may be no less disturbing to a patient than ones that are organically produced" (*Afterfects* 65). The mind/body dualism, though useful for preserving the humanist integrity of the soldier as a *cogito* encased in a damaged materiality from the implicitly feminized accounts of psychological disorder, disintegrated within the symptom of aphasia. Indeed, the most lasting contribution made by Head and Goldstein to aphasiology was their assertion that even though there may be a physical lesion in the brain, the aphasic symptom always appeared as the result of new integrations carried out by all available portions of the central nervous system. And for both, it was only by carefully listening to this articulation of disorder, attending to the complex narrative progression of the illness, that the story of the functional complexity of brain activity might appear in which the clinician could intervene therapeutically to become a protagonist, alongside the patient, in a tale of heroic overcoming.

Narrative interruptions

In 1919, Head published a volume of verse entitled *Destroyers, and Other Verses*, and his poetry reveals much of the sympathy with Romantic versions of human subjectivity and the resistance to mechanistic models that can be perceived in his medical writings. As befits a collection from a doctor who once prescribed Goethe as a form of therapy to a neurasthenic patient (Jacyna, *Medicine* 33), a large part of the volume was devoted to translations of the German Romantic poetry of Heinrich Heine; but he also published original verse clearly influenced by his experience of treating wounded soldiers. Too old to enlist, Head revealed again a certain equality between soldier and doctor, although in the poem 'I Cannot Stand and Wait', it is the doctor's sacrifice and work that needs to be elevated to be placed alongside the soldier's, in the hope that "when triumphal candles have burned low", "[p]erchance my gift may glow" (*Destroyers* 7). In the first verse of the poem, initially sent to fellow neurologist Charles Sherrington in 1915, he wrote:

How can I serve who am too old to fight?
I cannot stand and wait
With folded hands and lay me down at night

> In restless expectation of the day
> Will bring some stroke of Fate
> I cannot help to stay.
> Once, like a spider in his patterned web,
> Based on immutable law,
> Boldly I spun the strands of arduous thought,
> Now seeming naught,
> Rent in the sudden hurricane of war. (*Destroyers* 7)

The doctor, ennobled by his war work, finds himself drawn away from his previous position as one who categorized and ordered pathology towards a narrative of therapeutic heroism. The poem suggests, despite a persistent belief in the redemptive value of medicine, that the pre-war doctor who was able to sit "like a spider in his patterned web, / Based on immutable law", can no longer sustain his position. Instead, the boldly spun "strands of arduous thought" and the web, clearly analogous to neatly categorizable models of nerve systems such as those elicited from Head's earliest experiments in 1884 on reflexes in the lungs, or the maps of punctate cerebral function beloved of pre-war neurology, are destroyed by the shock of modernized conflict; they are "[re]nt in the sudden hurricane of war". A profound suspicion of mechanization emerges from the poems, as that which disturbs the continuity between humanity and the natural world. In 'Elan Vital', an organic, somewhat Bergsonian, account of a creative evolution in which "[a]ll things that live and grow are full of hope" appears in the face of the fact that "Death exacts a toll / From Beauty, Courage, innocent Desire"; but a certain, if poetically cloying, optimism remains. 'Elan Vital' affirms, in the end, the necessity of "stumbl[ing] on", "[i]n pain, in sorrow and bewilderment / Impelled to hope by man's instinctive soul" (*Destroyers* 15).

Jacyna has noted that although both Head and Goldstein displayed continuities with classical aphasiology, both are more easily integrated into the holistic movement that was gaining currency amongst neurologists in the immediate post-war period (*Lost* 149). There, brain function and "man's instinctive soul" spoke more consonantly to one another. In his work on sensation with Rivers in 1908, Head had already been keen to emphasize that the evolutionary history of the nervous system revealed "the means by which an imperfect organism has struggled towards improved functions and psychical unity" (qtd. Jacyna, *Medicine* 130), whilst Goldstein insistently asserted that the human being was an 'organism' rather than a machine, driven by a desire to achieve wholeness and coherence even in the face of its own anxious vulnerability to the assaults of an increasingly mechanized modernity. Although Goldstein had been a student of Wernicke's, his fascination with Romantic literature and idealist philosophy, alongside his collaboration with the Gestalt psychologist Aldhémar Gelb at the Institute in Frankfurt, led to the articulation of a new holistic neurology in tune with the Gestaltist revolt against both the chaos into which a world war had thrown Europe, and those seemingly unnatural systems of pseudo-order that read brain processes as technologized structures within the human machine.

Gestalt psychology placed itself in opposition to classical associationism, which was fundamentally derived from British empirical philosophy and conceived of the mind as a nonspatial 'space' in which basic units of mentation were joined into more complex systems according to immutable laws such as those of contiguity and similarity. As Anne Harrington puts it, "the whole process was seen to be automatic, determined, and curiously 'mindless'" (14). Harrington explains that "associationism held that all mental atoms were derivative but reliable packets of sensory data, variously defined ... as 'copies', 'pictures', or 'representative images' of direct experience. The concept of innate, inspired, or intuitive knowledge had no place in this model" (14), and the localizationists used this concept of a mechanically associating mind or areas of brain function to subtend their positivist assertions. Aligned with a late nineteenth- and early twentieth-century use of the term 'Gestalt' which is connected to Romanticism, holism and life, from the 1910s onwards Gestalt psychology attempted to reconfigure this mechanistic conception of the brain/mind relation, demonstrating empirically, using the apparatus of psychophysics and experimental psychology, that lived reality was not ordered according to the random display, generation and subsequent ordering of meaningless elements in a mechanically associating brain, or determined by the causal processes of a reflex arc. Instead, the human mind was constructed according to an *a priori* tendency to perceive the immanent structural wholeness that subtends life.[10]

The Gestalt psychology of the 1920s and 1930s was characterized by the urge towards the expression of wholeness in human personality, experience and abstract thought, and the perception of the external world as a symmetrical articulation of such internal holism. Goldstein translated this latent ordering into a theory of brain function, noting from his observation and treatment of wounded soldiers that the underlying and universal disturbance in *all* brain injury was a defect in the 'figure ground function' – that ability to pick out pattern and meaning,

Gestalts, from a background sea of phenomena (*Language* 5). With the brain-damaged patient, one may "expect the figure as reaction to a stimulus, and the patient may answer with the background"; the inverse also applied (*Language* 5). Extrapolating from this position, Goldstein argued that the immanent organization of the brain functioned to order the chaos of its experience into wholes, to effect a holistic symmetry between itself and its *Umwelt* (a term Goldstein borrowed from idealist philosophy and Jakob Johann von Uexküll). Language was deemed to pattern and synthesize experience in a similar way, permeating the totality of mental functioning. Language use, if it was non-pathological, then, allowed for the holistic activity of what Goldstein termed 'categorical' behaviour (*Organism* 45) – the ability to see the abstract category into which particular concrete events could be sorted.

The healthy individual thus had flexibility, the ability to maintain a sense of abstract categories, and a capacity to use language that was both propositional and volitional, enabling the perception and representation of certain Gestalts; and Goldstein argued that the brain-damaged patient would naturally compensate and re-orient brain function towards the expression of new methods of holistic organization. So although brain damage could be figured as an assault from modernity, with Goldstein also asserting that, "so-called nervous degeneration has a source in a mismatch between our capacities and the demands made on us by the progress of culture" (Goldstein, *Uber* 66; qtd and trans. Harrington 143), a new spirit of the age could nevertheless manifest itself through an increasing awareness of the brain's self-actualizing possibilities, its natural tendencies towards holistic organization and a rationality based upon the perception of latent Gestalts.

Significantly, for Goldstein, previous neurological theories of brain localization had misread their evidence of negation: they had taken the "terrible fateful step from the localization of symptoms...to the localization of function" (*Uber* 18; qtd. and trans. Harrington 152). Goldstein determined that because what was disturbed in brain damage was precisely the capacity to see the world holistically, the appearance of mechanization was, in fact, a product of the pathology exacerbated by the isolation of faculties during scientific observation and testing (*Organism* 50) rather than any manifestation of latent structures. Head also asserted that it was a mistake to imagine that language pathology simply revealed "the functions which combined for its development, exactly as a chemical salt can be analysed into an acid and a base" (*Aphasia* 477). Although Head's holism is less thorough-going than Goldstein's, he too insisted, using metaphors drawn from the realm of human social activity and nature rather than machines, that "the activity of the mind as a whole is uninterrupted; removal of one aptitude does not cause the collapse of the whole structure like a house of cards. The field of consciousness remains continuous as before; it closes over the gap as the sea and leaves no trace of a rock that has crumbled away" (*Aphasia* 494).

In terms of the clinical observation, diagnosis and treatment of aphasia, Goldstein was thus clear in 1927 that "one should never contemplate a phenomenon in isolation from the entire sick person... [for] Man is a psycho-physical organism. Each disease changes him in his entirety" (*Selected* 163–4). As a consequence, each patient needed to be assessed phenomenologically – a method which, according to the philosophical school with which Goldstein was aligning himself, demanded a 'return to the things themselves' without reference to pre-existing models against which behaviour might be mapped. Of course, a pre-existent tendency to synthesize the chaos of experience into wholes did pervade the functioning of the brain in this account, but the modalities by which the healthy organism effected such organization, and the damaged brain's adaptations and reorientations around lesions towards a new holism, presented themselves as complex, diachronically unfolding narratives which demanded a careful, involved attention and interpretation.

So the model of the brain as an adaptive organism, orientated towards a holistic organization, emerged, in both Head and Goldstein's work, from a certain phenomenologically inflected return to the patient. In 1903, Head had worried, in a letter to Ruth Mayhew, whom he would later marry, that his trenchant psychological anatomization of his mother, using attitudes drawn from his clinical work, had simply turned him into a "horrible recording machine" (qtd. Jacyna, *Medicine* 87). But he was later to render clearly, and in terms related to those of Kittler's 1900 discourse network, that clinical meaning emerged not simply as data impressed on the doctor as an objective observer, a "recording machine". Writing in 1919, Head asserted that the old diagnostic methods needed to be cast aside in order to match the fact that the War had "drawn an indelible mark across the history of our times" ('Disease' 365). Where Army medical officers had been happy to use the 'procrustean bed' of a *List of Diseases* to record and effect diagnoses, Head proposed a more active model of interpretation, of sifting meaning from the complexity of data recorded within the unfolding of the symptom. He advocated a new clinical practice that could integrate a modern and technologically inflected laboratory-based medicine with a holistic response to the patient's story of disease and recovery – a response that emerged from the doctor's application of his interpretative, aesthetic sensibility. "True diagnosis is an orderly procedure in which all the faculties of the mind, logical and instinctive, play their part", he wrote ('Disease' 366).

Recording the subjective narratives of patients' progressive adaptation to their condition – narratives that required the doctor to read them but whose shapes were also elicited from the patient himself as he was encouraged, through exercises and work, to assume a self-legible position within a broader social world – thus inserted an organismic diachrony into those socially isolated, spatialized and mechanized models of function so beloved of the late nineteenth century. By registering and reproducing the recursive, complex temporality of the organism, even as it was stabilized by the explicit teleology of an attempt to recover holistic organization, this apashiology determined a continuity between the brain-damaged and the healthy subject of modernity in which one was not simply the negative imprint of the other; rather, in Goldstein's terms, "pathological phenomena [could] be recognised as an indication of lawful variations of the normal life process" (Goldstein, *Organism* 30). Consciousness and representation, whether pathological or healthy, as precipitates of a holistically organized brain, were perceived to function according to processes of adaptive reaction, in which meaning appeared as it was sifted as a figure from ground. As Head put it, alluding to the new physiology of visuality later described by Crary, "the field of vision and the field of consciousness are built up on the same principles of ignoring parts devoid of sensation or reactions outside the focus of interest" (*Aphasia* 541). Both perception and meaning thus arose from processes of interpretation from which there emerged the idea of human subjects – both patients *and* doctors – as organisms actively thrown into the processes of their own becoming in the world.

More becoming than being

When offering his account of subjectivity as a complex, embodied being in the world, it is suggestive to note that Maurice Merleau-Ponty turned to the work of Goldstein and Gelb for both conceptual frameworks and empirical data.[11] In *Phenomenology of Perception* (1945), he critiqued a Cartesian tradition that seemed to turn the world into a collection of objects and subjects, and the human into a "body as the sum of its parts with no interior, and the soul as a being wholly present to itself without distance" (198). This Cartesian subject found itself nailed to a world dominated by "the transparency of an object with no secret recesses, the transparency of a subject which is nothing but what it thinks it is" (198). But for Merleau-Ponty phenomenology's method of 'reduction' enabled a certain parallax effect that allowed the complex ties and involvements of the phenomenal world to re-emerge. He critiqued Husserl's idea of reduction as being, at base, idealist, in its suggestion that one should bracket off the question of what existence objects of thought might have apart from the mind; for this phenomenological mode troublingly reduced the unexpected character of the experience of the world into simply the thought of the world. Merleau-Ponty nevertheless retained an etiolated conception of reduction as something that could effect a conceptual estrangement from the world, weakening our absorption in its seemingly limpid qualities:

> Reflection does not withdraw from the world towards the unity of consciousness as the world's basis; it steps back to watch the forms of transcendence fly up like sparks from a fire; it slackens the intentional threads which attach us to the world, and thus brings them to our notice; it alone is consciousness of the world because it reveals the world as strange and paradoxical. (*Phenomenology* xiii)

As with Heidegger's broken hammer, this reduction, as the etymology of the word suggests, became a *restoration* of the materiality of world and the subject, traced through with unexpected and unpredictable attachments.

For Merleau-Ponty, the experience of embodiment spoke against the idea that the subject might be able to exist in a world from which it had the structural capacity to separate itself as a discrete being. Instead, embodiment, once attended to, rendered explicit the processes through which the subject came to a sense of itself according to a narrative of experience emergent from the chiasmic intertwining of matter and idea, the natural and cultural, individual and social. For the body was

> rooted in nature at the very moment when it is transformed by cultural influences, never hermetically sealed and never left behind. Whether it is a question of another's body or my own, I have no means of knowing the human body other than that of living it, which means taking up on my own account the drama which is being played out, and losing myself in it. (*Phenomenology* 198)

In *The Structure of Behaviour* (1942), Merleau-Ponty had criticized the idea of body and mind functioning according to laws of mechanical function or association, refusing the notion that the reflex arc was the fundamental unit through which one could explain behaviour. Using the work of Goldstein, he argued instead that the human organism functioned by orientating itself towards gestalts, although, significantly, these were not part of the world *in itself*; rather, they appeared from the organism's constitution of its existence as a sensing and conscious being whose experiences were meaningful to it (*Structure* 143; *Phenomenology* 60). For Merleau-Ponty, the body was experienced by a mind that emerged in a continuum with its sensations and perceptions; mind and body were, in turn, chiasmically entwined with, and constituted by, a world that was rendered legible by being formed as a vital milieu in which the embodied subject was involved.

Because Merleau-Ponty refused any dualism of mind and body, subject and object, even representation – thought and language, in this case – emerged as part of the totality of the organism's experience and its imbrication within the world of sensation. Representation, however, was also that which enabled such intersections and interactions with the world to become meaningful. Extrapolating from Goldstein's account of the complex losses of orientation towards the world experienced in aphasic disturbance, he asserted that language did not express thoughts that somehow pre-existed or outlasted the experience of material embodment: rather, language "brings home to us the enigmatic nature of our own body" (*Phenomenology* 197) as implicated in a process of creating and interpreting meaning. For, using terms drawn from neurological paradigms, like language itself, the body

> is not a collection of particles, each one remaining in itself, nor yet a network of processes defined once and for all – it is not where it is or what it is – since we see it secreting in itself a 'significance' that comes to it from nowhere, projecting that significance on to its material surrounding, and communicating it to other embodied subjects. (*Phenomenology* 197)

The stimuli of the world were not simply alien shocks, fashioned and controlled (or not) by the organism; instead, it was precisely through the power of representation, of giving significance, that the embodied subject understood itself as orientated towards a world of possibility and engagement from which it drew and to which it imbued significance. He went on to note the paucity of accounts of language anterior to Goldstein's holistically reconfigured account:

It has always been observed that speech or gesture transfigure the body, but no more was said on the subject than that they develop or disclose another power, that of thought or soul. The fact was overlooked that, in order to express it, the body must in the last analysis become the thought or intention that it signifies for us.... This disclosure of an immanent or incipient significance in the living body extends...to the whole sensible world, and our gaze, prompted by the experience of our own body, will discover in all other 'objects' the miracle of expression. (*Phenomenology* 197)

Aphasiology had illuminated the continuity between pathological and healthy linguistic function, but it had also revealed that a neurologically inflected sense of being-in-the-world, to use Heidegger's phrase, inaugurated a subject that was neither split between its immaterial and material nature nor simply a mechanically associating machine; rather, this linguistic subject of modernity was fundamentally embodied through and according to its own capacity to represent.

Citing both Head and Goldstein, Merleau-Ponty wrote: "Whether the authors are aware of it or not, they are trying to formulate what we shall call an existential theory of aphasia, that is, a theory which treats thought and objective language as two manifestations of that fundamental activity whereby man projects himself towards a 'world' " (*Phenomenology* 190–1).[12] He found in them, but also learned from them, that an embodied subject, a neurological subject of modernity, was one whose thought, language and corporeality always emerged from middle of things, from a complex narrative of translation and interpretation which always began *in media res*. Despite its strong materialism, classical aphasiology and theories of localization of brain function had found themselves retreating from this complexity, in epistemological terms, by imagining transcendental, scientifically pristine modes of knowledge that could turn the brain into a functional machine to be apprehended according to a spatially stable model of vision. But as the sounds and narratives of the aphasic patient persisted and insisted and were let back into the clinical writings of aphasiology, there emerged new models of language and perception which forced a reformulation of meaning as that which emerged from a continuum of chiasmic intertwinings of the natural and the cultural, the somatic and the psychological. Language itself had been reconfigured, in alignment with the new writing technologies of the period, and by both experimental psychology and aphasiology, as something that appeared through the work of interpretation from the sometimes assaulting spray and swell of internal and external phenomena; language surfaced only through processes which, according to the tendency to sort information into gestalts, could separate figure from ground, meaning from the flux and excess of the sensation and mentation. And this new model of language both enabled and extended the discursive conditions from which a Romantic, holistic, yet historically urgent, neurology could develop within the post-war period. Still bound to a strong materialism in its use of clinical data, its reproduction of the complex materiality of the experience of their patients, and its explicit insistence that the doctor, patient and the wider social sphere were bound together in mutually constitutive ways, this aphasiology became central to new modes of understanding the embodied and the thinking subject within whom the capacity for language and abstract thought emerged from and indeed determined a chiasmic implication with the world. From the broken hammer of disturbed language, a whole world of involvements thus appeared as a figure from the occulting ground of normative experience. This work of translation and interpretation rather than apprehension in the production of meaning that brain-damaged patients illuminated, in their disabilities and in the narratives through which they restored legibility to themselves, thus offered up a new paradigm for understanding both an emergent, embodied-thinking subject of modernity and a world that neither transcended nor could simply be reduced to thought.

Notes

1. In this phrase, usually translated as 'rational animal', Heidegger reminds his readers of the strong connection between *logos* as rationality and language. For him, *zoon logon ekon* becomes "that living thing whose Being is essentially determined by the potentiality for discourse" (47).
2. See Arsleff for an account of these debates.
3. In 1864 Armand Trousseau published 'On aphasia, a sickness formerly wrongly referred to as aphemia', where he criticized Broca's term with (pseudo-) philological arguments (*aphemia* could imply 'infamy'). He proposed the term 'aphasia' (from the Greek meaning 'without language') instead (Tesak and Code 54).
4. It is suggestive that although Leborgne 'always answered *tan, tan*' to any questions, Broca chooses not to read the utterance as 'tantan', or 'tonton', which means 'uncle' (47).
5. Head notes: "it was possible to estimate with considerable accuracy the extent of their education, and the ability with which they had carried out the more exacting of their military duties" (*Aphasia* 148).
6. Goldstein explicitly links his establishment of hospitals in which programs for retraining could be implemented with those organized in England by Head (*Aftereffects* 67).

7. In 1919, Prussian law asserted that occupational counselling and employment agencies should be established with the assistance of psychologists (Rabinbach 278). Goldstein's Frankfurt Institute was able to return 90 per cent of its patients to employment after treatment (Harrington 146).

8. See Head's 'clock tests' (*Aphasia* 155–6), and the account of Case No. 10's difficulties with summer time (194).

9. Goldstein asserted: "although it is not always easy to exclude neurotic reactions, experience reveals that men with *real* injuries of the brain and prominent symptoms, such as ... aphasia ... seldom show neurotic reactions" (*Aftereffects* 219, emphasis added).

10. Harrington notes that because Gestalt theory was empirically orientated, socially liberal and dominated by Jewish intellectuals, it cannot be aligned with the conservative romantic holism of the burgeoning Nazi ideology, against which it explicitly opposed itself. Max Wertheimer believed that the logic to be detected in the way in which minds spontaneously worked could be marshalled to teach people how to think in 'structural truths', according to 'Gestalt logic', thus protecting universalist principles against mechanized or relativistic thinking (133).

11. Merleau-Ponty organized a French translation of Goldstein's *Der Aufbau des Organismus* to be published as the second volume of the *Bibliothèque de Philosophie* series of phenomenological works, following Husserl's *Ideas for a Pure Phenomenology* (Harrington 158).

12. It should be noted that Head and Goldstein both demurred from offering such an existential account as both, in the end, were committed to positing an ontological position as anterior to any personal one. For Goldstein, this can be traced to the Kantian preoccupations of his youth.

Works cited

'Aphasia.' *Oxford English Dictionary*. 2nd ed. Oxford: Clarendon, 1989.

Arsleff, Hans. *From Locke to Saussure: Essays on the Study of Language and Intellectual History*. London: Athlone, 1982.

Broca, Paul. 'Notes on the Site of the Faculty of Articulated Language, followed by an Observation of Aphemia.' 1861. *Reader in the History of Aphasia: From Franz Gall to Norman Geschwind*. Ed. Paul Eling. Amsterdam: John Benjamins, 1994. 41–5.

Clarke, Edwin, and L. S. Jacyna. *Nineteenth-Century Origins of Neuroscientific Concepts*. Berkeley: California UP, 1987.

Crary, Jonathan. *Suspensions of Perception: Attention, Spectacle, and Modern Culture*. Cambridge: MIT Press, 2001.

Descartes, René. *The Philosophical Writings of Descartes: Volume III, The Correspondence*. Trans. John Cottingham et al. Cambridge: Cambridge UP, 1991.

Goldstein, Kurt. *Aftereffects of Brain Injuries in War*. 1942. New York: Grune & Stratton, 1948.

——. *Language and Language Disturbances: Aphasic Symptom Complexes and their Significance for Medicine and Theory of Language*. New York: Grune & Stratton, 1948.

——. *The Organism: A Holistic Approach to Biology Derived from Pathological Data in Man*. 1934. New York: Zone, 1995.

——. *Selected Papers/Augewählte Schriften*. The Hague: Nijhoff, 1971.

——. *Uber Rassenhygiene*. Berlin: Julius Springer, 1913.

Harrington, Anne. *Reenchanted Science: Holism in German Culture from Wilhelm II to Hitler*. Princeton: Princeton UP, 1996.

Head, Henry. *Aphasia and Kindred Disorders of Speech*. Vol. 1. London: Cambridge UP, 1926.

——. 'Disease and Diagnosis.' *British Medical Journal* 1 (1919): 365–7.

——. *Destroyers and Other Verses*. London: Humphrey Milford, 1919.

——. 'Hughlings Jackson on Aphasia and Kindred Affections of Speech.' *Brain* 38 (1915): 1–27.

Heidegger, Martin. *Being and Time*. Trans. John Macquarie and Edward Robinson. Oxford: Blackwell, 1962.

Jacyna, L. S. *Lost Words: Narratives of Language and the Brain: 1825–1926*. Princeton: Princeton UP, 2000.

——. *Medicine and Modernism: A Biography of Sir Henry Head*. London: Pickering & Chatto, 2008.

Kittler, Friedrich. *Discourse Networks: 1800/1900*. Trans. Michael Metteer and Chris Cullens. Stanford: Stanford UP, 1990.

——. *Gramophone, Film, Typewriter*. Trans. Geoffrey Winthrop-Young and Michael Wutz. Stanford: Stanford UP, 1999.

Lichtheim, Ludwig. 'On Aphasia.' *Brain* 8 (1885): 433–84.

Merleau-Ponty, Maurice. *Phenomenology of Perception*. Trans. Colin Smith. London: Routledge, 1962.

——. *The Structure of Behaviour*. Trans. A. L. Fisher. Boston: Beacon, 1963.

Rabinbach, Anson. *The Human Motor: Energy, Fatigue and the Origins of Modernity*. Berkeley: California UP, 1992.

Richardson, Alan. *British Romanticism and the Science of the Mind*. Cambridge: Cambridge UP, 2001

Tesak, Juergen, and Chris Code. *Milestones in the History of Aphasia: Theories and Protagonists*. Hove: Psychology Press, 2008.

Wernicke, Carl. 'Some New Studies on Aphasia.' 1885. *Reader in the History of Aphasia: From Franz Gall to Norman Geschwind*. Ed. Paul Eling. Amsterdam: John Benjamins, 1994. 90–8.

——. 'The Motor Speech Path and the Relation of Aphasia to Anarthia.' 1884. *Wernicke's Works on Aphasia: A Sourcebook and Review*. Ed. Gerturde H. Eggert. The Hague: Mouten, 1977.

Young, Robert M. *Mind, Brain and Adaptation in the Nineteenth Century: Cerebral Localization and its Biological Context from Gall to Ferrier*. Oxford: Oxford UP, 1990.

11
Shell Shock as a Self-Inflicted Wound, 1915–1921

Jessica Meyer

On 7 November, 2006, the British government announced that it would issue pardons to the 306 British soldiers executed for military offences during the First World War. This marked the culmination of a long-running campaign for pardons led by the Shot at Dawn organization. According to this association, most of those "brutally gunned down by the authorities" were "clearly suffering from shell-shock" (*Shot at Dawn* 3). As an example, the organization's webpage details the case of Harry Farr, who had been diagnosed and treated, apparently successfully, for shell shock, six weeks prior to his execution. While Farr's case has gained considerable notoriety, having been the subject of a court case seeking his pardon brought by his daughter against the Ministry of Defence, the link between shell shock and execution is by no means as clear in all 306 cases (Todman).[1] The popular view, however, is that "many of those shot for cowardice were suffering from post-traumatic stress disorder after enduring months of artillery bombardment in the trenches" ('300 WWI soldiers receive pardons' para. 1). As Gertrude Harris, Farr's daughter commented on hearing of the government's intentions, "I have always argued that my father's refusal to rejoin the front line . . . was in fact a result of shell shock, and I believe that many other soldiers suffered from this, not just my father" ('Victory for 93 year old' para. 1).

This association between shell shock and the pardoning of those shot for military offences including cowardice is based on the assumption that shell shock is a medical condition. This interpretation serves to extenuate the disciplinary infractions men suffering from the condition may have committed, an interpretation that is reinforced by the BBC's use of the term 'post-traumatic stress disorder' (PTSD) to define Farr's condition ('300 WWI soldiers receive pardons' para. 1). PTSD is not, in fact, the same as shell shock. While both are general terms that embrace a wide variety of physical and mental symptoms for which no organic cause was readily apparent, shell shock was initially defined by Charles Myers as "the result of functional dissociation arising from the loss of the highest controlling mental functions" (51). During the war years the term 'shell shock' rapidly lost status as a diagnostic term within the medical profession. Instead, doctors used a wide variety of apparently more specific classifications, including hysteria, neurasthenia and 'Soldier's Heart', even as the phrase 'shell shock' entered popular discourse (Meyer 198). By comparison, post-traumatic stress disorder is an acknowledged psychological illness, listed in the American Psychiatric Association's *Diagnostic and Statistical Manual (DSM-III)*, giving it, as Ben Shephard has pointed out, "a real intellectual authority" that was never accorded to shell shock. As a medical condition, PTSD is cited as the explanation for aberrant behaviours, excusing such behaviours because their cause is deemed to be medical rather than the result of moral failings. To read shell shock in this manner is, however, a recent development in a debate over the relationship between psychological disorders and military discipline that has existed since before the term 'shell shock' was coined in 1915 (Young 39).

Throughout the First World War, and well into the interwar period, the range of physical and psychological symptoms that came collectively to be known as 'shell shock' were the subject of intense debate because they fell uncomfortably between the care of medicine and the strictures of military discipline. Lack of clear organic cause of the symptoms made it difficult to classify them as a medical condition, despite the efforts of neurologists such as F.W. Mott to uncover physical symptoms such as lesions on the central nervous system through brain dissection (Shephard 30). At the same time, the loss of self-control that characterized all the symptoms, including hysterical paralyses, contractures, nightmares, hallucinations and insomnia (Meyer 198–9), challenged any easy assumption that the sufferers' aberrant and undisciplined behaviour was intentional.

Only one other form of illness or injury needed to negotiate competing discourses of military discipline and medicine during the First World War: the self-inflicted wound. Like shell shock, self-inflicted wounds served to undermine military authority through the deployment of medical symptoms of ambiguous causation. Men who shot themselves created wounds that needed medical attention and care but also removed themselves from the danger of death that fulfilling their military duty imposed. These wounds thus raised questions about policing, medicalization and the nature of self-control that mirrored those raised by shell shock. This chapter examines the status of both shell shock and self-inflicted wounds in medical and military discourses to explore how the conflicts they caused between discipline and medicine affected understandings of self-hood, the body and control during and after the First World War.

Shell shock

The term 'shell shock' was coined by Dr Charles Myers in 1915 as a term to describe a variety of physical symptoms, which were becoming increasingly common among British soldiers and for which there was no obvious organic cause. One early theory for this behaviour, espoused by Sir Frederick Mott, the leading neurologist at Maudsley Hospital during the war, was that the blast from artillery shells, the dominant weapon of the war, was creating microscopic lesions on men's nerves, causing their otherwise apparently inexplicable symptoms (441–9). This theory gained considerable currency among both neurologists and the wider public, to the extent that it "embedded itself, in a crude and over-simplified way, in the public imagination" (Shephard 28). This physical explanation for the otherwise seemingly inexplicable behaviour com-bined with Myers' evocative and alliterative term to create an idea of a medical illness as the root cause for the rash unmilitary behaviour being witnessed by 1915. By 1916, however, Myers had renounced both the relatively crude neurological explanation of the condition and the term 'shell shock' in favour of a distinction between 'concussion' and 'ner-vous shock' which encompassed an interpretation of symptoms caused as much by the general conditions of warfare, including exhaustion and horror, as by specific moments of physical danger. Unlike Mott, who viewed non-physical explanations for shell shock symptoms as evidence of "an acquired or inherited neuropathy" (448), Myers spent much of the war advocating treatment for men immediately behind the front line in order to return them to duty more successfully and rapidly. He believed that the causes of shell shock were more complex than any inherent lack of soldierly qualities, arguing that "[b]etween wilful cowardice, contributory negligence (i.e. want of effort against loss of self-control) and total irresponsibility for the shock, every stage and condition of shell-shock may be found" (qtd. in Shephard 49).

Myers considered himself a psychologist rather than a neurologist, and bitterly resented the 'invasion' of the field of shell shock by more straightforward neurologists such as Gordon Holmes, a specialist on the physiology of the brain, who was given the control of shell shock cases on the north of the Somme front in December 1916, effectively ousting Myers (Shephard 47–8). While Myers took sick leave in England following his demotion from the position of Consulting Psychologist to the Armed Forces, the influence of his psychological theories persisted. Most famously advocated by Myers' Cambridge colleague Dr. W.H.R. Rivers in his 1921 book *Instinct and the Unconscious*, this theory argued that shell shock was, in fact, two forms of neurosis arising from the conditions created by war, principally the "conflict between the instinct of self-preservation and certain social standards of thought and conduct, according to which fear and its expression are regarded as reprehensible" (208). One solution to this psychic conflict was "the occurrence of some disability ... which, so long as it exists incapacitates the patient from further participation in warfare and thus removes all immediate necessity for conflict between instinct and duty" (207). This reaction, which manifested itself in physical or hysterical symptoms, was what Freud, whose theories influenced those of Rivers, termed a "flight" into illness (Freud 3). In anxiety neuroses, by contrast, the instinct for self-preservation was similarly dominant but resulted in an increased conflict with a highly developed sense of duty that led to the more psychological symptoms of nightmares and hallucinations. However, as with the physical disabilities and overt emotional expression that marked the hysteric for most doctors (Shephard 9), the depressive symptoms of the anxiety neurosis also, according to Rivers, arose out of the conflict between the unconscious instinct for self-preservation and the conscious understanding of duty. The primary difference, in Rivers' argument, was that the hysterical patient, usually a private, defined duty in terms of the following of orders, making such patients highly suggestible and thus prone developing conversion symptoms; the neurasthenic, however, was usually an officer who viewed duty in terms of responsibility and thus tended to experience symptoms of anxiety (Rivers 217–9).[2]

This interpretation of the causes of shell shock highlights the problems that shell shock posed for the military establishment which relied on a socially enforced sense of duty to persuade men to place themselves actively in harm's way. In the view of the British Army "[m]en were either sick, well, wounded or mad; anyone neither sick, wounded, nor mad but nonetheless unwilling to or incapable of fighting was necessarily a coward, to be shot if necessary" (Shephard 25). While the physical explanation of the condition put forward by Mott and others placed shell shock firmly within this narrative of behaviour, the impossibility of proving the existence of lesions of the nervous system hampered attempts to either cure or prevent the condition. By 1916 the "wastage" of manpower (Shephard 45) in the army had reached levels that were unacceptable to the military authorities in practical terms, forcing them to turn to alternative explanations of the condition. At the same time, many within the military hierarchy continued to view the symptoms of the conditions as indicative of cowardice rather than of an illness or wound (Shephard 45–6).

The psychological reading of the condition was even more problematic for military authorities. The notion of the flight into illness, the withdrawing of willingness or ability to do one's duty as a soldier, appeared to support the view that the condition was simply a form of cowardice. Yet the instinctive and therefore unconscious nature of this withdrawal, as argued by Rivers (6), complicated such simple classifications. Physical symptoms such as tremors, blindness and paralysis – many of them terrifying and painful in themselves – made it more difficult to classify shell shock as an easy way out of service even if a discharge on the grounds of unfitness was the ultimate result. They thus more closely resembled a form of wounding, although a wounding of the mind rather than of the body. This fact was to some extent acknowledged by the military in the form of an attempt to distinguish between shell shock caused by wounds and those deemed to be a sickness resulting from the sufferer's physical, mental or moral failings rather than from enemy action. The former, which involved a letter W prefix on the victim's casualty report and entitled him to wear a wound stripe and attain the status of 'wounded', was diagnosed in cases where shock was due to 'enemy action'. All other cases were labelled with a letter S and the victim did not attain the honourable status of the wounded soldier (Shephard 28–9). Such distinctions were considered important, as Major W.J. Adie later explained to the Committee on Enquiry into 'Shell-Shock': "My opinion is that no man who had simply broken down mentally should be given a wound stripe, but the man with an obvious commotional shock who has been buried or blown up deserves one. I distinguish rather sharply between the two conditions" (qtd. in Southborough et al. 17). To classify disabilities resulting from a perceived lack of self-control alongside those that could be identified as resulting from physical force, as doctors such as Rivers and Myers were doing, was to blur the distinction made by Adie by acknowledging all forms of shock as a form of psychic wound. It was the expanding of the definition of wounds to include the functional alongside the physical that troubled men like Adie, upsetting as it did their notions of soldierly behaviour which encompassed an individual's control over both his body and mind. The blame for such functional injuries, without clear organic cause or structural change to the body, could be said to lie with the self rather than an outside enemy; so giving them the status of wounds challenged the military training which condemned self-inflicted wounds for deliberately removing soldiers from their duty.

The situation was further complicated for the military authorities by the temptations that shell shock appeared to present to the malingerer if acknowledged as a genuine medical condition. Malingering, defined by Joanna Bourke as a particular form of shirking that centred on the body (Bourke 81), was a common practice among soldiers throughout the war. Indeed, it was a generally acknowledged and even communal phenomenon, with soldiers hawking specimens of tuberculosis-infected saliva to their comrades and obliging medical orderlies deliberately creating diseases for men, sometimes for a fee (Bourke 84–5). Shell shock, with its lack of clear organic causation and the breadth of symptoms it could encompass, had great potential to be used by malingerers as a conscious flight from military service. Lieutenant-Colonel G. Scott-Jackson, for one, believed that "[i]t was only when it began to get into the papers and the men came back home and found the amount of sympathy at home for the man 'shell shocked' that his battalion got it" (qtd. in Southborough et al. 46).[3] Unsurprisingly, front-line medical officers, who were primarily responsible for policing malingering (Bourke 89–94), were among the most sceptical and stringent in their attitudes towards those claiming to be suffering from the illness, an attitude reflected in the incident related by Max Plowman when he asked a medical officer to view an unconscious soldier:

> The doctor bends over him for a moment, and then, rising, shouts with astonishing fury: 'You damned young scrimshanker, get up! What the devil do you fancy you're playing at? Think you can swing the lead on me? Get up, or I'll have you in the guard-room.'
> He pushes the boy with his foot, but the lad does not stir.
> 'Don't you think he is ill?'
> 'Ill! There is nothing the matter with him at all. Just 'wind up,' the bloody young coward. Leave him there if he doesn't get up, and don't call me again. I don't waste my time over these damned scrimshankers.'
> He turns and goes back to the dug-out. (Plowman 56–7)

J.C. Dunn, a medical officer attached to the Second Battalion of the Royal Welch Fusiliers where he met Siegfried Sassoon, was equally sceptical towards shell shock. Upon hearing of Sassoon's transfer to

Craiglockhart Military Hospital, he wrote: "Sassoon's quixotic outburst has been quenched in a 'shell-shock' retreat. He will be among degenerates, drinkers, malingerers, and common mental cases, as well as the overstrained" (372). Dunn also believed that most others diagnosed with shell shock were equally undeserving, even those labelled like Sassoon as 'overstrained'. Charles Moran, a medical officer with the Royal Fusiliers, acknowledged the existence of shell shock as a condition but believed that it derived from hereditary weakness as much as genuine illness:

> When men had been at the mercy of their instincts, when they had given up the effort to control fear and had abandoned themselves to their impulses, there was something in their make-up that made them less willing or less able to behave unselfishly. It was not fear, many who had been most miserably afraid had done splendid things. It was not evil chance of war time or place, the hellish bombardment, or war would be a lottery in which the unlucky drew the stamp of cowards. No: it was only bad stock that brought defeat (26).

This view was echoed by W.J. Adie who believed that "many of us were suffering more or less from 'shell shock', which made us not so efficient, and yet we remained in the line...[but] all sorts of people got out of the line with so-called 'shell shock', and the result was that they evaded their full duty and were not punished" (qtd. in Southborough et al. 17). For the men who were on the front line of policing soldiers' health, shell shock was often viewed as a moral problem, whether ascribed to hereditary weakness of will or the cowardly instinct of the malingerer, rather than a medical condition.

Self-Inflicted wounds

Shell shock was not, however, the most serious form of malingering that medical officers were required to police. Far more dramatic were the cases of self-mutilation in which men either shot themselves in the hand or foot, or got comrades to do it for them. Alternatively, men could tempt a shot from the enemy that would earn them a 'blighty' wound and a trip home (Bourke 84). Such behaviour was clearly deemed to be malingering and the punishment for those who were caught was a term of imprisonment (General Staff, War Office 223) – the fate of the 3,894 men convicted of causing self-inflicted wounds during the war.

As a type of malingering, self-inflicted wounds tended to be viewed far more harshly by the military authorities than other forms. Evidence for this can be seen in the severity with which it was treated by the military justice system. As Helen McCartney has pointed out in relation the Liverpool Territorials, "[i]n all but two cases... [self-inflicted wounds] warranted a court martial" (172). By comparison, "[i]t was taken for granted that men would report sick without cause in order to gain a few days' respite from the line. This kind of short-term malingering appears to have been condoned by the Battalion hierarchy, as long as the men were not permanently lost to the unit" (171–2). The distinction being made was clearly the permanence of the malingering. Short-term respite could be countenanced, particularly at times when the battalion was not over-stretched. A man who wounded himself, however, as opposed to imagining a temporary illness, was "rejecting his role as an infantryman, and attempting to remove himself *permanently* from the war" (McCartney 172).

While such actions might be viewed harshly by the military authorities as evidence of a failure on the part of a soldier to fulfil his duty, it was harder to cast the actions of those with self-inflicted wounds as those of a coward. As Guy Chapman wrote of one Private Turnbull, "[p]erhaps those who call this man a coward will consider the desperation to which he was driven, to place his rifle against the foot, and drive through the bones and flesh the smashing metal" (122). The distinction being made concerns the level of bodily self-control required to create the wounds. Shell shock was increasingly defined by the loss of conscious control over the body to the unconscious instinct for self-preservation. As David Drummond wrote in a letter to the editor of the *Lancet* in 1918, "loss of control... is, of course, the essence of the condition" (349). While this reaction was viewed as a genuine medical condition by many doctors specializing in the treatment of mental disorders, it could also be viewed as a form of cowardice by some, as noted by Professor G. Roussy of the Faculté de Médicine de Paris:

> Cowardice, I consider is a lack of self-control of the individual in the presence of a situation in which there is an element of fear. Any man who can control himself is a courageous man, but the man who runs away, or who does certain other actions not esteemed worthy, is defined a coward. A courageous man is a man who can exert his self-control. (qtd. in Southborough et al. 21)

By contrast, the self-inflicted wound, while it was motivated by a similarly 'cowardly' desire to escape the fearful circumstance in which men found themselves, required a temporary *conscious* rejection of the instinct for self-preservation and necessitated extreme physical self-control for the flight into disability to be achieved. The consciousness of the subversion of military training and medical authority was thus much more threatening and required clearer condemnation and sanction in the form of arrest and court martial.

Yet not all self-inflicted wounds required a high level of conscious decision in their creation as not all were self-administered. In a number of cases, men clearly created risks in order to invite wounds or even death, such as J.W. Graystone, who volunteered for the dangerous duty of manning a listening post in the hope of receiving a light wound which would excuse him from front-line duty (Graystone, 17 April 1916). Nor were such actions always so conscious. R.G. Dixon went through a "lengthy spell of to-hell-with-it-all-what-does-it-matter, and in the popular phrase of a later time, couldn't care less.... This particular spell lasted about six weeks, I remember, during which time I behaved like a complete idiot and did the silliest things and, though I had plenty of narrow shaves, I never got hit" (20). Dixon clearly was not conscious of wanting to remove himself from the war through wounds or death, yet his actions in placing himself in harm's way might have had that result.

Situations like this blurred the lines between the conscious self-inflicted wound and the unconscious flight into illness of shell shock, in which symptoms manifested themselves without clear organic cause. The cause of any wounds received from reckless action could be clear, but the motivation behind the behaviour that invited them was often opaque, even to the man whose actions resulted in his receiving a wound. Acts of apparent bravery were thus open to question. Siegfried Sassoon, for example, described the actions that saw him awarded a bar to his Military Cross as the result of "over-strained nerves [which] had wrought me to such a pitch of excitement that I was ready for any suicidal exploit" (445).

There was a clear awareness among front-line soldiers of the potential ambiguity in the definitions of their own and others' actions as the results of cowardice or shell shock. It was the level of apparent consciousness that ultimately defined the levels of sympathy with which both conditions were viewed by such men. CQMS Gordon Fisher acknowledged, there was such a thing as "genuine shell-shock" (qtd. in Macdonald 476). For the ordinary soldier, this condition could be pitied rather than condemned, just as the actions of Private Turnbull in shooting himself in the foot were pitied by his superior officer who hoped he would receive a light sentence, but who nonetheless viewed a trial and a criminal sentence as a necessary outcome arising from the commission of a military crime.

Treatment

Shell shock and self-inflicted wounds thus both occupied a liminal space between medical disability and military offence, demanding both treatment and censure, although the question of in what proportion remained hotly contested. This contestation created problems for the military establishment in its attempts to maintain authority. Self-inflicted wounds were to be punished, where identified, requiring doctors to act as detectives as they did when investigating shell shock as a possible form of malingering. George Coppard, for example, recalled being treated with suspicion and hostility by the nursing staff at a casualty clearing station when a wound to his foot caused by the accidental discharge of a comrade's revolver was suspected of being self-inflicted (Coppard 102–3). In addition to being medically treated as a victim of a wound, the sufferer of a self-inflicted wound was also treated as a criminal whose conscious action in causing the wound had to be tried and punished. That the maximum penalty for this crime was imprisonment rather than death reflects the view of military authorities that desertion, of which the self-inflicted wound might be viewed as a form, was less severe an offence than cowardice, a classification which, as we have seen, was more difficult to apply (Oram 38). Where desertion could be defined by an act of wilfully not being in the place one was ordered to be by the military authorities, the judgement of what acts defined cowardice, ranging from refusing to advance to surrendering to throwing down arms and running away, was more subjective.

In cases of shell shock, which could appear as more cowardly in that sufferers avoided the dangers of the front line at apparently little physical cost, the investigation of the condition as a form of malingering could be as harsh. In addition, the ability of the military authorities to condemn the actions of the sufferer was further complicated by the role of the unconscious in causing the behaviour of which the conscious individual could be viewed as a victim. Because of this proximity to cowardice, discipline continued to be viewed as an essential element in the reduction of cases of shell shock. It was not simply that, as Oram has argued, the death penalty was used by the armed forces as a way of eliminating "worthless" men, among them shell shock victims who were viewed as insufficiently military (84–101). As the 1921 War Office Committee of Enquiry into 'Shell-Shock' noted, "the panic stricken individual in general recovers under the influence of discipline and those qualities which make for moral courage, though it was on record that in isolated cases men who at first showed evidence of cowardice, fought bravely and regained confidence in themselves when the alternative of death stared them in the face" (Southborough et al. 94–5). Thus discipline, including the threat of death, retained a strong influence in political and military theories of treating both those overcome by temporary panic and those who initially showed evidence of cowardice in the years after the war.

As Ted Bogacz has noted, the Committee of Enquiry's conclusions show "evidence of the committee's struggle to reconcile the modern ambiguous notion of shell-shock with traditional absolutist norms for behaviour in war and peace" (248). While attempting to define absolute values of courage and cowardice, the committee did acknowledge that shell shock was "a reversion to the instincts, and may be communicated from permanently unstable minds to momentarily unbalanced minds" (Southborough et al. 94).Where instinct – the aspect of men's nature that was associated with animalistic behaviours ('The Neuroses of War' 71), rather than consciousness, involving the control of such instincts through the application of willpower – prevailed, the easy distinctions between cowardice, physical courage and duty could no longer be as effectively maintained. Just as self-inflicted wounds demonstrated a level of physical courage even as they exposed the victim/perpetrator's cowardly instinct to escape the dangers of military service, so the instinctive nature of shell shock served the dual purpose of excusing the victim from culpability for his actions at the same time as it marked him as less than soldierly for his failure of self-control.

To further complicate matters, the Committee of Enquiry attempted to maintain the status quo with regard to questions of causation and treatment of shell shock, at the same time acknowledging the developments in both theory and practice that had occurred during four years of war. Thus the report argued that shell shock might occur in those momentarily unbalanced but it was more likely to be seen in those who were already and permanently unstable. And, while medical treatment was necessary for those who were fearful but not cowards, discipline remained an important element of the medical approach to those deemed to be lacking the moral quality of courage as "cowardice [could only] be regarded as a military crime to be punished when necessary by death" (Southborough et al. 139). By 1921 shell shock had come to occupy space at a nexus between medical care and military discipline that it shared with the self-inflicted wound. Indeed, the conclusions of the committee come close to defining shell shock in terms that are a reminder of the status of the self-inflicted wound as a condition that had to be simultaneously treated and punished in the attempt to return the sufferer to his military role. The ambiguities inherent in the diagnosis of the physical damage of the self-inflicted wound thus influenced military and medical approaches to the psychological damage of shell shock, even as the complexities that arose in the treatment of shell shock shaped professional attitudes towards the self-inflicted wound.

Conclusion

Shell shock and self-inflicted wounds can be seen, therefore, to be very similar conditions. Both involved a flight into disability as an escape from the fears and horrors of warfare. Both appeared to subvert military discipline and medical authority. Both could elicit some pity and understanding from fellow soldiers. Both were treated with a combination of medical care and military discipline in order to return sufferers to military usefulness as well as preventing further cases.

The most significant difference between the two conditions was in the role of consciousness in creating the disability. The introduction of the unconscious as an element of causation of shell shock was resisted by many medical professionals and complicated the position of the military and political establishments in their treatment of the condition. The unconscious, after all, could not be punished for its actions in the way that the conscious individual could, while to punish the sufferer with no conscious control over his actions appeared unjust.

Yet in the acceptance by an establishment body such as the Southborough Committee of Enquiry of a place for the unconscious in the causation of shell shock, we can see a process of normalization of a problematic war trauma that mirrors that of the public embracing of the term itself. The association of a physical explanation for the condition with a new phrase, whose emotive power is evident from its rapid and permanent adoption and adaptation by popular culture, combined conservative thinking about psychological disorders with the new dominance of artillery in warfare – a development in arms associated with the modernity of the First World War (Sheffield, *Forgotten Victory* 274–6). Similarly, when the concept of the unconscious was applied to the pre-existing idea of the self-inflicted wound, a less rigid and more sympathetic view of the sufferer could be taken. This was only one of a number of changes the mass recruitment of civilians into the armed forces made to the status of British military authority (Barham

116; Sheffield, *Leadership* 151) While psychological disorders and their treatments had existed prior to the First World War, it was only through this marriage of tradition and modernity that they were able to enter the cultural imagination and discourse. It would take another world war, however, before such disorders were removed from the sphere of military discipline to become the modern medical diagnoses that shape contemporary discourses of appropriate military behaviour.

Every attempt has been made to obtain copyright permission for unpublished works quoted.

Notes

1. As Gary Sheffield has noted, the blanket pardon does not "distinguish between those who deliberately let down their country and their comrades and those who were not guilty of desertion of cowardice" (qtd. in '300 WWI soldiers receive pardons'). Several cases included in the pardon involved clear military misdemeanours, including striking officers and deliberate desertion (Todman para. 10).
2. This distinction was also made by Rivers' mentor, the neurologist Henry Head, in his evidence to the War Office Committee on 'Shell-Shock' (Shephard 57–8).
3. There is, in fact, little evidence for this opinion as there are no figures indicating the distribution of occurrences of shell shock within the armed forces. There was certainly, however, what Peter Barham calls "a significant counterblast of 'civilianness', in which the citizenry succeeded not necessarily in thwarting, but certainly in restricting the dominion of militarized social relations over disadvantaged groups such as the psychiatric casualties of war" (Barham 116). This counterblast tended to find its expression, however, in lobbying for access to treatment and pensions rather than expressions of sympathy concerning the conditions from which men actually suffered.

Works cited

'300 WWI soldiers receive pardons.' *BBC News*. Accessed 12 Jan 2007. <http://www.news.bbc.co.uk/1/hi/uk/4796579.stm>

Barham, Peter. *Forgotten Lunatics of the Great War*. New Haven: Yale UP, 2004.

Bogazc, Ted. 'War Neurosis and Cultural Change in England, 1914–22: The Work of the War Office Committee of Enquiry into 'Shell-Shock'.' *The Journal of Contemporary History* 24.2 (April 1989).

Bourke, Joanna. *Dismembering the Male: Men's bodies, Britain and the Great War*. London: Reaktion, 1996.

Chapman, Guy. *A Passionate Prodigality: Fragments of an Autobiography*. London: Ivor Nicholson & Watson, 1933.

Coppard, George. *With a Machine Gun to Cambrai*. London: Papermac, 1986.

Dixon, R.G. 'The Wheels of Darkness.' Ts. memoir. c.1970s. Imperial War Museum, London.

Dunn, J.C. *The War the Infantry Knew 1914–1919*. 1938. London: Abacus, 1994.

Drummond, David. 'Letters to the Editor: War Psycho-Neurosis.' *Lancet*. 2 March 1918. 349.

Freud, Sigmund. 'Introduction.' *Psycho-Analysis and the War Neuroses*. Ed. S. Ferenczi et al. London: The International Psycho-Analytic Press, 1921.

General Staff, War Office. *Field Service Pocket Book (1914)*. London: His Majesty's Stationary Office, 1914.

Graystone, J.W. T. Diaries. 91/3/1. Imperial War Museum, London.

McCartney, Helen. *Citizen Soldiers: The Liverpool Territorials in the First World War*. Cambridge: Cambridge UP, 2005.

Macdonald, Lyn. *1915: Death of Innocence*. London: Headline, 1993.

Meyer, Jessica. "Gladder to Be Going Out Than Afraid': Shellshock and Heroic Masculinity in Britain, 1914–1919.' *Uncovered Fields: Perspectives in First World War Studies*. Ed Jenny Macleod and Pierre Purseigle. Leiden: Brill, 2004.

Moran, Charles. *The Anatomy of Courage*. London: Constable, 1945.

Mott, Frederick. 'Lettsomian Lectures on the Effects of High Explosive on the Central Nervous System.' *Lancet*. 26 February 1916. 441–9.

Myers, C.S. 'A Final Contribution to the Study of Shell Shock.' *Lancet*. 11 January 1919. 51–4.

Oram, Gerard. *Worthless Men: Race, eugenics and the death penalty in the British Army during the First World War*. London: Francis Boutle, 1998.

Plowman, Max. [Mark VII]. *A Subaltern on the Somme in 1916*. London: J.M. Dent, 1927.

Rivers, W.H.R. *Instinct and the Unconscious: A contribution to the biological theory of the psycho-neuroses*. Cambridge: Cambridge UP, 1922.

Sassoon, Siegfried. *The Complete Memoirs of George Sherston*. 1937. London: Faber & Faber, 1952.

Sheffield, Gary. *Forgotten Victory: The First World War: Myths and Realities*. London: Review, 2002.

——. *Leadership in the Trenches: Officer-Man Relations, Morale and Discipline in the British Army in the Era of the First World War*. Basingstoke: Macmillan, 2000.

Shephard, Ben. *A War of Nerves: Soldiers and Psychiatrists 1914–1994*. London: Pimlico, 2002.

Shot at Dawn Campaign. Accessed 17 January 2007. <http://www.shotatdawn.org/page3>.

Lord Southborough et. al. *Report of the War Office Committee of Enquiry into 'Shell-Shock'*. London: HMSO, 1922.

Todman, Dan. 'A step too Farr?' *Trench Fever*. Accessed 13 January 2007. <http://www.trenchfever.wordpress.com/2006/08/16/a-step-too-farr/>

'Victory for 93 year old as her father, Private Harry Farr, receives a pardon for cowardice: Posthumous human rights victory for Harry Farr.' *IrwinMitchell*. Accessed 13 Jan 2007. <http://www.irwinmitchell.com/Victoryfor93yearoldasherfatherreceivesapardon.htm>.

Young, Allan. *The Harmony of Illusions: Inventing Post-Traumatic Stress Disorder*. Princeton: Princeton UP, 1995.

12
Modernity and the Peristaltic Subject

Jean Walton

There is no dearth of literature on the cultural symbolism of shit; indeed, scatological studies of waste reveal much about how we distinguish the clean from the impure, the self from the abject, the sacred from the profane. But what if we were to shift the focus from the end product of digestion to what I am calling, in a larger project, the modern 'peristaltic subject', considered from the perspective of its dynamic, material implication in comprehensive systems of movement, flow, consumption, production and elimination? The very temporality of modernity moves to the foreground in such a study, permitting us to ask not what the body produces, but how it experiences, and deploys, *time*. It helps us to see how we govern the movement of substances through our own bodies, even as we ourselves move through larger networks of goods, money, information, pleasures and affect. It allows us, too, to shed light on an aspect of what Foucault, in *Discipline and Punish*, calls the accumulation and capitalization of time "in the body" that has not yet been adequately theorized: in other words, on the visceral training of the body as a "processor" of the world, not only in terms of our cognitive participation in symbolic exchange, but also in terms of our unconscious implication in disciplinary structures (157).

To pursue a study like this, we must turn to a neurological account of the peristaltic system that dates back to about the same era in which scientific managers applied time and motion studies to the labouring body, to increase efficiency and decrease waste in both the industrial and the domestic workplaces.[1] In this account, the peristaltic system is sometimes under the domain of the brain and central nervous system, but more often controlled by what has come to be termed 'the second brain', a division of the autonomic nervous system that completely escapes directives from the first brain, and yet which is responsive (and therefore susceptible) to the natural and cultural environment. In this account, our gut 'thinks by itself' insofar as it functions independently of our first brain; but at the same time, like the leg that marches, the arm that aims the rifle, and the hand that practices perfect penmanship, it offers another target for the modern fashioning of the 'docile body'. Mary Ann Doane has commented that modernity "was characterized by the impulse to *wear* time, to append it to the body so that the watch became a kind of prosthetic device extending the capacity of the body to measure time" (4); I would argue, in fact, that the body itself became, and continues to function as, a kind of chronometer, its neurological capacities engaged for the purpose of gauging and modulating duration, thus making of itself a guarantor of both the regularity and the uni-directionality of time.

In the interest of brevity, I will limit myself to two representative medical accounts of how digestion and excretion work, one by gastroneuroenterologist Michael Gershon, whose 1998 book *The Second Brain* offers a compelling account of the enteric nervous system, the other by a less known medical practitioner, Josiah Oldfield, whose 1943 article on constipation was typical of the discourse of regularity throughout the modern era. Read together, these texts offer a striking Darwinian account of the human body in an evolutionary context, and demonstrate that while the neurological model of the 'second brain' was developed in the early twentieth century, when the discursive distinction between organic and psychological concepts of the 'nerve' had not yet solidified, it has since received renewed attention, at the turn of the twenty-first century, allowing for a particularly labile approach both to the embodiedness of linguistic and cultural aspects of subjectivity, and to the social aspects of embodiedness. As we will see, the modern concept of the nerve and the persistent history of its ambiguity as at once a physiological, a 'mental', and even a cultural phenomenon, *especially* with regard to its role in peristaltic processes, permits a particularly promising approach to the question of the relationship between the subject and the social realm.

Gershon begins with the conventional understanding that our bodies are, for the most part, structured hierarchically when it comes to nerves. We have a central nervous system, consisting of the brain and the spinal cord, and a peripheral nervous system, which takes orders from the central processing headquarters. In addition to this, the nerves which connect the brain and the spinal cord to the rest of the body, and which carry the commands sent out by the central controlling organs, may be classified under two broad categories: the skeletal motor system and the autonomic nervous system.[2] Generally (though there are exceptions to this rule), the skeletal nerves require our conscious attention to carry messages to our muscles (to set the table and salt our potatoes, for instance), while the autonomic nervous system is involuntary, carrying orders from our brain that escape our conscious attention, but which are essential to internal bodily functions. The autonomic nervous system can be further classed into two major subdivisions, depending on which aspects of the body are being serviced: the sympathetic division, which pertains to responses involving the entire body, such as heartbeat and blood pressure, and the parasympathetic division, which is restricted to single organs like the bladder or the pupils (14).

Gershon does not depart from this conventional model of the human nervous system until he arrives at the scientific insight that gives the intriguing title to his book. Turning to a 1921 text entitled *The Autonomic Nervous System*, he notes that most physicians seem unaware that J.N. Langley, the author of this text, divided the autonomic system not into two (sympathetic and parasympathetic), but into *three* categories. Western medicine seemed to have forgotten or ignored Langley's discovery of this third division of nerves since the publication of his book, at least until Gershon drew renewed attention to it in the late twentieth century. This emerging, disappearing and resurfacing division of nerves is of paramount importance for anyone interested in the vagaries of peristalsis, because its duties are entirely limited to the operation of the intestines, and for this reason, Langley dubbed it, in this text, the 'enteric nervous system'.

What distinguishes the enteric nervous system from the other two divisions of the autonomic system is that it doesn't seem to be connected to the central nervous system. Whereas the sympathetic and parasympathetic nerves receive and act upon messages from the first brain, apparently the enteric nervous system can make decisions all on its own, without taking orders from a higher authority. Indeed, Gershon describes the autonomy of the enteric nervous system in terms of contrasting systems of governmentality in the body. The central nervous system "clearly outranks the peripheral nervous system," and thus gives the orders. "Commands flow *from* the brain and spinal cord via the nerves of the peripheral nervous system *to* muscles and glands (the effectors of the body).... The brain is at the top, the effectors and sensory receptors are at the bottom" (16–17, emphasis in original). But in a section headed 'Rebellion in the Bowel: The Brain Below', Gershon announces an insurrection against this otherwise hierarchical arrangement. Since the enteric system can "process data its sensory receptors pick up all by themselves, and ... can act on the basis of those data to activate a set of effectors that it alone controls," it may be considered a kind of "rebel, the only element of the peripheral nervous system that can elect *not* to do the bidding of the brain or spinal cord. ... The enteric nervous system is thus an independent site of neural integration and processing. This is what makes it the second brain" (17, emphasis in original). Unlike all the other organs in the body, whose reflex actions cease when their connections to the central nervous system are severed, the peristaltic reflex of the gut can function as though it has a brain of its own.

Moreover, it almost seems as though the enteric brain is 'second' only in the sense that its discovery followed that of the 'first' brain; in the final analysis, both 'brains' are essential to the functioning not only of the body, but of culture in a larger sense. "The enteric nervous system may never compose syllogisms, write poetry, or engage in Socratic dialogue," Gershon comments, "but it is a brain nevertheless" (17). Another way to put this, perhaps, is to say that it is not the brain, but rather the mind, that engages in linguistic and cognitive activities, but that we risk losing the cerebral organ if our enteric brain functions badly. And yet, it is precisely the slippage between 'mind' and 'brain' that Gershon relies on to maintain the striking effect of his argument. While we think with our first brain, there is some more 'thinking' going on in our gut, a thinking of which we have not been aware. For Gershon, this unconscious thinking lies beyond the purview of the psychoanalyst, but is newly accessible to the biologist, armed as he is with the "chemical language" with which nerves "talk" to each other (25). To understand the enteric brain, thus, one must understand the functioning of serotonin and its receptors, since, as Gershon's research has proven, serotonin is one of the primary substances in this chemical language. I should note here that for Gershon, the use of terms like 'brain', 'language', 'talking' and so on is entirely metaphorical, serving only to illustrate processes that are purely biological. He does not mean to imply that the gut can actually participate in a symbolic system of communication or thinking, in the way that the mind (as opposed to the brain) can. Indeed, his argument becomes most polemical when it counters the widespread accepted wisdom among doctors and laity alike that so many enteric dysfunctions are 'psychological' in origin. The discovery of a virus as the cause of many ulcers, for instance, has revolutionized the cure for these ulcers, shifting attention from ineffective psychotherapeutic treatment of stress to pharmaceutical remedies that eliminate the ulcer's organic pathogen. But, as I will be suggesting, retaining a more literal understanding of there being a 'second brain' in the gut offers possibilities for understanding ways in which the gut communicates with the world, ways in which it is 'social' after all.[3]

Now, with regard to peristalsis, what, precisely, are our two brains doing? To answer this question, we must understand how Gershon, and his predecessor Oldfield, conceptualize the body, first, as it morphologically inhabits space, and second, as it gauges time via the regulation of movement through space. First, the morphology.

It is self-evident to say that the most obvious boundary between self and world is the continuous, protective organ that surrounds our interior and keeps us intact: our skin. Gershon, however, disrupts this model of the body as a self-evidently enclosed object in space with a more complex image, elaborated through the aid of a literalized poetic trope:

> The design of the body can be understood by paraphrasing T.S. Eliot. We are indeed hollow men and . . . women. The space enclosed within the wall of the bowel, its *lumen*, is part of the outside world. The open tube that begins at the mouth ends at the anus. Paradoxical as it may seem, the gut is a tunnel that permits the exterior to run right through us. Whatever is in the lumen of the gut is thus actually outside of our bodies, no matter how counterintuitive that seems. (84)

Gershon's model of the body as a hollow tube through which the world passes is not a new one, of course; any medical practitioner who focused on alimentary systems in the twentieth century would have proceeded on the assumption of such a model, among them Josiah Oldfield, whose article on constipation offers an exemplary version of this trope.

In one of several contributions to a special issue of the *Medical Press and Circular* devoted to 'Constipation: the Universal Malady' (1943), Oldfield applied himself quite methodically to the problem of how to remove obstacles to elimination in the human body. But to understand how constipation arose as a threat to the health and well-being of the human subject, he offered an elaborate account of the slow transformation, over evolutionary time, of an amorphous organism to that which, like Gershon's, was most notably characterized by its tube-like quality. We began as "amoebas," whose "selection of substances from the environment," "absorption of them," and "rejection of the remainder as being unwanted or unusable" was rather haphazard and indiscriminate

(381). The world simply moved into and out of us with no particular directionality. But "in common with those processes of evolution by which living organisms have become more complex, and their separate portions more specialised," we began to develop into organisms that consisted of "a tube running the length of the body. The food is introduced at one end. Selection, transformation and absorption of suitable elements from the mass takes place as the food passes along the tube, and then from the other end are cast out the waste, the unsuitable and the deleterious elements" (382). Now the world began to run through us with a particular directionality, entering at one end of the tube, and exiting at the other end. Moreover, in order to ensure that this movement occurred "in rhythmic and uninterrupted flow," an "elaborate system of nerves and muscles have been provided along the whole of this tube" (382).

The concept of the evolving body here is an interesting one, consisting most importantly of a tube, through which substances pass, and around which a network of mechanisms accretes, designed for the purpose of keeping the substances flowing through the tube. Oldfield pictured us as sites for 'processing' the world which was propelled through us as it in turn made us and sustained us. Moreover, to aid in illustrating peristaltic action, Oldfield borrowed the image of the 'ordinary garden worm' to give an inside-out demonstration of how we, as tubes, kept the world proceeding through us. Just as the worm moved through the earth through the aid of a "set of ripples running down the length of his body and pushing against the ground as they ripple," so, too, "the ripples along the inner muscular coating of the intestines continue for awhile and then stop – and the intestinal contents stop. Each time the ripple runs along it carries the contents a little way forward" (382).

Having described the body as it inhabited and processed 'space', Oldfield then introduced the element of time, or rather, the means by which the concept of 'duration' became broken down into component parts, as the organism became more sophisticated. In our bodies as earthworm-like tubes, the "ripple does not go right down to the end, otherwise the human being would be constantly oozing out faeces" (382). The oozing was prevented by the mechanism of a "lock gate near the end, called the sigmoid flexure," resulting in a "gradual heaping up of an increasing mass." The contents of the body tube waited in abeyance until a "push button set of nerves" was triggered, forcing "the whole collected mass clean out of the tube, and behold what is called 'a motion' has been accomplished" (382).

That we were not "constantly oozing out faeces" was apparently what set us apart from the lower species, according to this account of the human body.

> In the lower rungs of life this process of expulsion is so frequent as to be almost continuous. Whereas the higher we go up the more we find that the storage near the extremity gets larger in amount and the room provided for this storage becomes more dilated and consequently the emptying takes place at longer intervals. (382)

Unlike caterpillars, mice and cattle, who all seem characterized by "more or less continuous defaecation," humans made an important discovery which enabled them to introduce the concept of periodic accumulation into the flow which was animal life.

> When man discovered the value of time, and the manifold varieties of interests that it unfolded to him, he...concentrated his feeding times into two or three hours or less, in the twenty-four. This freed his propulsory processes from continuous action and therefore eating, propulsion of foods along his alimentary canal, transformation, absorption, and expulsion, became alternated with long periods in which mental activities replaced the previous continual devotion to...bodily nutrition. (382)

By this logic, an adult, according to Oldfield, had the impulse to eat only at intervals, and "by the over-riding action of mentality upon the nervous system a regular sequence habit can be developed so as to ensure a complete freedom from an impulse urge, excepting at times that can be arranged for ahead – such, for example, as after breakfast or before going to bed!" (382) The exclamation point, rare in Oldfield's text, seemed to emphasize that we had arrived at nothing less than the juncture of the invention of history, born at the moment schizzes were introduced into the natural flow of substances through the peristaltic tube that defined the organism as a body. Time now could be measured by the amount of substance that was stored up in the end of the peristaltic tube, kept in reserve for an evacuation that would happen only periodically, instead of continuously. Now time was characterized by periodicity, flows interrupted by schizzes, rather than an indistinguishable stream that allowed for no consciousness of one moment as opposed to another. As soon as this happened, a future could be imagined, and something planned for; the human subject now became capable of imposing will on the passage of time itself by means of a holding in reservation, then a releasing at a prescribed time. An 'impulse urge' might now be escaped insofar as it could be "arranged for ahead" – making it no longer, after all, an impulse, since it could be counted among the known events that would occur according to the day's schedule ("after breakfast or before going to bed").

The real problem Oldfield set out to address, of course, was not the 'constant oozing' that might be imagined if the peristaltic system worked without interruption, but rather the apparently widespread malady of constipation, when the contents of the bowel were stockpiled for too long, and failed to make their way out the other end. While Oldfield mentioned, in passing, "nervous irregularity in controlling muscular action" as a possible cause of intestinal stasis, the main culprit was our susceptibility to "fashion, habit, and novel ideas" about what constituted good nutrition (383). We failed to consume the proper food in order to ensure that the contents of our bowels were of adequate 'quality' and 'mass'. Spicy foods, for instance, were "irritants to the delicate nerve filaments along the colon, and thus injure their sensitiveness, so that in time they become inactive"; moreover, foods with too little 'mass' made for watery or meagre contents, giving the contracting and expanding intestinal walls nothing to grab onto, and thus "the contents of the intestine provide no fulcrum for an onward push by the muscles. The contractile ring action in the intestine may, therefore, as easily push fluid content backward as forward" (383–4).

It was up to us to provide the necessary fulcrum by making sure that we selected just exactly the correct part of the world to ingest, and thus to direct on its way through ourselves. Otherwise, the movement of the world through us might either reverse in direction, or come to a stop. This would happen if the world in us were either too soft or too hard; in both cases, peristaltic motion would fail, and nothing would appear at the other end of the tube. Meaning, perhaps, that *we* would stop as well. The reversible worm image functioned doubly. It gave us our model for understanding internal peristaltic action in the human body, of course. But because of its reversibility, its susceptibility to being imagined inside out as well as backwards and forwards, it also carried with it the image of our own bodies as they moved through the peristaltic systems of our world of "fashion, habit, and novel ideas."

Yet even when diet is ideal, neither Oldfield nor Gershon seem prepared or willing to address the vexing question of what happens when the 'push button mechanism' that releases the contents of the bowel fails to function. What exactly is meant for instance by Oldfield's obscure passing reference to "nervous irregularity in controlling muscular action"? What set of nerves is meant here? Who would be exerting the control? To approach this question, we return to Gershon's account of the division of nerves involved at each stage of the peristaltic journey, from mouth to anus. The reader is positioned as the traveller in this section, and invited to imagine herself journeying through a long passageway, coming, as she presses onwards, now under the jurisdiction of a central sovereign authority (first brain), now under the control of local authorities (second brain), depending on which division of the nervous system is in play at a given point in time and space.

"In every body," Gershon begins, "the brain is king. Its writ is law. At the top of the bowel, the rule of the king is acknowledged, but as one descends deeper and deeper into the depths of the gut, the rule of the king weakens. A new order emerges: that of the second brain" (113). From mouth to stomach, the sympathetic and parasympathetic systems are involved, taking their orders from the brain. But

> to descend below the *pyloric sphincter* (the exit of the stomach)... is to move almost beyond the reach of the king. This is the turf of the enteric nervous system where the brain can exert only quantitative effects and not make the basic decisions of what to do and when. It is an autonomous region that cares little for the brain and is happy to do without it altogether. (113, emphasis in original)

It is as though one vanishes from all conscious and unconscious cognizance of the brain while in the depths of the intestines; this is a curious territory, not 'in' the body proper, of course, because still separated from it by the lumen of the intestine. This is a place where the world passes through us, and becomes in some sense 'unknown' to our central nervous system, while at the same time being under the control of our 'second brain', a brain which *is* inside us, transforming the world moving through us while we are unawares.

But where the world returns to 'itself' as it were, emerging from the end of the tube which is our bodies, the first brain takes over again, almost with a vengeance. As one passes from the colon to the rectum and the anus, the ambiguity about who or what is in control becomes particularly pronounced. For here, a quite complex set of conditions must come about before 'a motion' is accomplished. It is not as simple as Oldfield's 'push button mechanism' makes it sound, and even Gershon, whose real passion is the second, not the first brain, can give only a cursory account of the vicissitudes of the defecatory impulse.

So what happens when the contents of the colon reach their end-point, and the question of how they are to be expelled arises? For anyone who has noticed that only certain social, habitual, environmental, or emotional conditions must be present before they can properly experience the urge to eliminate (and act upon the urge), this may be the most crucial part of the peristaltic system. "As the rectum fills," Gershon tells us, "it delivers a 'call to stool.' This sensation is usually not painful, although it may become so as it rises to a crescendo; nevertheless, the 'call to stool' is not a sensation that is easy to ignore. When felt, it tends to lead to action sooner rather than later" (171). If the reader experiences some anxiety that we might spew the noxious contents of our intestines into the world, Gershon reassures us not to worry, that "gravity is not a threat to our social well-being" since our rectum is "well endowed anatomically to handle the heft and bulk of the stool it contains." To maintain the "veneer of civilization" the rectum is cunningly designed to permit "storage of stool" for "limited periods of time," and its exit is guarded not by one, but two tightly locked doors: the internal and external sphincters (171).

So who is addressed by the rectum's interpellation, its "call to stool?" At first, it would seem to be these two sphincters that hold in the rectum's contents: they are summoned, and they open in response to this summoning. But no, Gershon says, "these sphincters...are not slaves that open as soon as the 'call to stool' is issued. They can hold stool back and resist opening until the moment is right. In fact, a guiding principle upon which the smooth working of our society is based is that the anal sphincters of every individual who is not an infant will indeed hold stool back until the right moment" (171). The "smooth working" of society, it appears, depends on the temporary blockage of the "smooth working" of peristalsis.

So now the question arises as to who or what determines when the "right moment" is, thus permitting the sphincters, which have been putting up an oppositional resistance to the second brain's propulsive mechanisms, to open and allow expulsion of the rectum's contents. By the time we reach this last stage in the peristaltic process, the second brain has ceded most, though not all, responsibility once more to the first brain which, depending on information gathered as to the state of the social environment external to the body, authorizes the sphincters to block or not to block. The perceptual and cognitive systems, passing information to the central nervous system, inform it that the "right moment" has come, and the central nervous system in turn permits the sphincters to open, allowing for the final stage of peristalsis to be completed.

But, as we may have suspected, it is not as simple as that. Consider, after all, that there is a question not of one, but two sphincters. Most of us are aware, of course, of the *external anal sphincter*, which is made up of skeletal muscle and is under our volitional control. We can decide whether to relax it or not. The internal sphincter is regulated however "only by the enteric and autonomic nervous systems" and is thus "on autopilot and out of our volitional control" (171). Before the external sphincter even gets the chance to retain or release the contents of the bowel, based on the conscious signals we send it from the first brain, this internal sphincter has to have already opened, has to have already made a decision to respond to the "call to stool" sent out by the full rectum. If this inner sphincter does not open for some reason, the outer one may relax all it wants, but nothing will pass through. Here is where the question of constipation, or the inability to pass stool, becomes most perplexing. For, on what basis does this *internal* sphincter decide that the "right moment" has arrived to respond to the call to stool? It falls under the jurisdiction in part of the enteric nervous system, which Gershon would have us believe is oblivious to what's going on in the social world in which the body must conduct itself, and the autonomic nervous system which, though connected to the first brain, operates independently of our conscious deliberation. Here, presumably, is where the question of the "right moment", indeed, the very question of temporality, is taken over by unconscious mechanisms; where the conscious mind, detecting the "call to stool," positions the outer sphincter over a perfectly acceptable toilet, gives the order to let go and, in some cases, nothing happens. "Regularity" is far from being an uncomplicated autonomic function; nor is it strictly, I am arguing, a matter of individual dietary control, contrary to what Oldfield, Gershon, and medicine in general would suggest.

In the 'Docile Bodies' section of *Discipline and Punish*, Michel Foucault turns to the school, the barracks, and the manufactory to note the emergence of

an important phenomenon: the development, in the classical period, of a new technique for taking charge of time of individual existences; for regulating the relations of time, bodies and forces; for assuring an accumulation of duration; and for turning to ever-increased profit or use the movement of passing time. (157)

"How can one capitalize the time of individuals," Foucault asks, "accumulate it in each of them, in their bodies, in their forces or in their abilities, in a way that is susceptible of use and control?" The disciplines, he argues, accomplish the individualising of humans through the training of the body to become a "machinery for adding up and capitalizing time" (157). Turning a Foucauldian eye to the modern peristaltic body reminds us that even before the subject ever enters the school, the barracks, or the factory, s/he undergoes the incremental disciplinary techniques of the nursery, the kitchen, and the bathroom, and is subjected to the meticulous training not only of the voluntary skeletal nervous system, but also of the autonomic enteric nervous system.

In the first decades of the twentieth century, the young subject, as a consumer of nutrients and producer of waste, was expected practically from birth, through a series of graduated exercises, to discipline the peristaltic system of his/her body – or rather, to *be* disciplined via the training of the lower intestine, bowel, and sphincter muscles. Toilet training involved nothing less than the mastery of the bodily regulation of time: the division of duration into evenly spaced evacuations of the large intestine, the measuring of progress by rhythmic contractions of smooth muscle, and the guarantee of a distinction between the 'natural time' of the indiscriminately excreting animal and the 'cultural time' of the systematically excreting human being. The properly trained peristaltic system also served to confirm the uni-directionality of time, as evidenced by the proper transformation of food into fuel and waste, and the distance maintained between point of ingress (the mouth) and point of egress (the anus).

Two temporalities were involved in this process: the cyclical time of ingestion and expulsion, and the developmental time of the disciplining of the viscera. While the world was detained and released, detained and released, an accumulation of a different kind occurred – not that of faeces in the large bowel, but of a habitude that was stored in the musculature, nerves, and viscera of the organism: what was stored was the ability to store in the first place, a capacity to hold in reserve before expelling, a talent for discerning the prescribed place and time. If time was thus 'capitalized' in the body, it was through the agency not only of the first, but of the second brain and its dealings with the world.

Moreover, these dealings were as much about affective states of being as about physiological processes: Brian Massumi, for instance, classes the enteric nervous system as the 'deepest' among three systems of perception through which we take in, and act upon, sensory data. There are the 'exteroceptive senses' (sight, hearing, tactility, for instance) which make direct contact with objects in the world; this is followed by proprioception, which "translates the exertions and ease of the body's encounters with objects into a muscular memory of relationality. This is the cumulative memory of skill, habit, posture" (59). Finally, there is the "interoceptive sense," which also occurs at the "dimension of the flesh" where it

> intervenes between the subject and the object. It, too, involves a cellular memory and has a mode of perception proper to it: *viscerality* (interoception). Visceral sensibility immediately registers excitations gathered by the five "exteroceptive" senses even before they are fully processed by the brain. (60, emphasis in original)

Massumi acknowledges Gershon's rediscovery of the 'second brain' as the basis of his speculations about the enteric nervous system, which, as he puts it, "provides one of the physiological bases for the autonomy of affect" that drives the overall argument of his book (266n13). He offers a mundane example of how the interoceptive perceptual apparatus of the gut registers an event in the external world:

> As you cross a busy noonday street, your stomach turns somersaults before you consciously hear and identify the sound of screeching brakes that careens toward you.... The immediacy of visceral perception is so radical that it can be said without exaggeration to precede the exteroceptive sense perception. It anticipates the translation of the sight or sound or touch perception into something recognizable associated with an identifiable object. (60–1)

The enteric nervous system, Massumi says, "empirically describes one of the ways in which our body thinks with pure feeling before it acts thinkingly" (266n13). Moreover, though his example has to do with the "functioning of 'viscerality' in relation to shock," he argues that stress, "which might be thought of as slow-motion shock" is also processed by and through the enteric nervous system.[4] I would add that we need not even posit the events registered, and "thought," by the visceral perceptual apparatus in terms of their function as disruptive pathogens (shock, stress) to imagine this condition in which our "body thinks with pure feeling before it acts thinkingly," especially if we keep in mind the routine ways that we target the interoceptive senses, from the first changing of the diapers, through the disciplinary techniques of toilet training, to the designation of certain architectural spaces and their furnishings as proper locales for the relaxation of inner and outer sphincter, to the segregation of these spaces depending on the genitalia of the one who is to defecate (the men's room vs. the ladies' room), and to the availability of these spaces depending on whether one's skin is perceived by the exteroceptive senses of one's fellows as white or not. I write here of the early twentieth century in the United States, England, and Europe, but of course, while the variations change, the fact of visceral training, and its implications for how we 'think with pure feeling' is in effect at any historical moment.

A brief foray into the vast literature on the 'habits' of infancy and childhood in the early twentieth century gives an instructive perspective on how the interoceptive senses were not only targeted, but also correlated with other affective and cognitive functions of the body. In the United States and Britain, almost every publication on childcare from this period (ranging from self-help books marketed to parents, to inexpensive government manuals, to articles in medical journals) focused specifically on the habits of eating, sleeping and elimination. The extent to which the training of the bowels was understood to enhance the child's integration within the larger social sphere, indeed, within the nation, is perhaps best exemplified in one of the three books published as the result of a White House Conference on 'Child Health and Protection' in 1930. John E. Anderson's *Happy Childhood* opened with a quotation from none other than the President himself, Herbert Hoover, who announced that the conference had gathered together all the best representatives of science to explore and understand the "safeguards and services to childhood which can be provided by the community, the State, or the Nation – all of which are beyond the reach of the individual parent," but who hastened to reassure us that "every scientific fact . . . every public safeguard . . . every edifice for education or training or hospitalization or play" were "but a tithe of the physical, moral, and spiritual gifts which motherhood gives and home confers" (Anderson vii). Where parenthood left off, the state intervened; by the same token, the state, with its science, could never provide what the presumably magical maternal instinct would accomplish. But, as the prescription for proper potty training unfolds in the texts of this period, we find that the well-intentioned instincts of "motherhood" functioned more often than not as impediments to the proper visceral training of the infant and child – hence the necessity to interpellate parents as agents of the 'home' insofar as this home was conceived as the primary locus in which the citizen was fashioned.

As with most of the literature published throughout the 1920s, the chapter on bowel training in *Happy Childhood* presumed a progression from an inchoate state of the body to one that had learned to measure and portion time appropriately. "In the new-born," we are reminded,

> bowel movements and urination take place at irregular periods, whenever the need is felt. In order to take a place in society the child must establish control of these processes, i.e., must learn to withstand pressure in the bladder and intestine when necessary and to meet his needs on appropriate occasion. The openings of both the bladder and the intestines are muscular structures under nervous control and are subject to training. (Anderson 59)

The ultimate objective was to reduce the habit of elimination to "the mechanical level"; indeed, the book was entitled *Happy Childhood* precisely because the very sense of happiness was presumed to be directly premised on the regularization of otherwise irregular processes:

> Happy is the child who masters these basic responses in his early years and moves on from the orderly healthful regime established by his parents to face with confidence the new situations presented by his ever-changing environment. Unhappy is that child who finds that the building of basic habits has been done in such a way that he must build up control anew in a period that should be free for an attack on habits appropriate to a higher level of development. (Anderson 68)

In another publication, Anderson offered much more specific advice about the best timetable for the training of the bowels, as well as detailed instructions for how and when the infant's body should be positioned in relation to certain facilitative objects in the world, and how best to suppress or mitigate the transmission of one's own affect during the course of the training process:

> Training may begin as early as the sixth week, if the baby is in good physical condition. The mother may hold the infant over a receptacle in her lap, the child's back being toward her, and supported by her arms and body. Warming the vessel or having a little hot water in it sometimes induces a movement more quickly. An enamel spittoon with broad, shelving sides is suitable for the baby's use at this period. (Anderson and Faegre 136)

A clock-like regularity should characterize this earliest period: "in beginning training, the same place and hour should be chosen each day, preferably directly after one of the morning feedings"; indeed, "unvarying regularity of hour, place, and conditions" was advocated, as was "removing distractions" and arranging "some sort of reward for the child when he is successful" (Anderson and Faegre 137).

When the pressure of a vessel against the child's buttocks was not enough to stimulate peristaltic movement at regular intervals on a daily basis, then it could be helpful "to use a suppository of soap or glycerine to alter the movement to a suitable time of day" (Anderson and Faegre 138). Indeed, the recommendation of the soap stick as a stimulant to the sphincter reflex was ubiquitous throughout the literature of toilet training, and was most vividly described in the following passage from another government publication that advocated "training the bowels" as early as the "end of the first month":

> Hold the baby in your lap or lay him on a table with his head toward your left, in the position for changing his diaper. Lift the feet with the left hand and with the right insert a soap stick into the rectum to act as a suppository. Still holding the feet up, press a small chamber gently against the buttocks with the right hand and hold it there until the stool is passed. The first time the soap stick is used the stool will come in 5 or 10 minutes. Later this time will be shortened. (Children's Bureau 59)

Note, in the accompanying illustration reproduced in Figure 6, the presence of the clock and knife used to fashion the soap stick next to it.

However, even the most assiduous adherence to a regular schedule would fail to systematize the baby's bowels if the mother or nurse did not also learn to properly calibrate the affective atmosphere that accompanied the training process. More particularly, a regimen of repression was to be established, both with regard to the mother's affective states, and then with regard to the child's. "The parent's attitude should show that she expects success by the child" wrote one author, "but under all circumstances irritation, scolding or punishment should be rigorously suppressed by the parent to avoid associating in the mind of the child this unpleasant accompaniment with his attempt" (Blatz and Bott 134). While the irregularity of the bowel movements in infancy were to be replaced by clock-like punctuality, the infantile pleasure experienced by the defecating body should be retained and protected from the pernicious influences of negative parental sentiments:

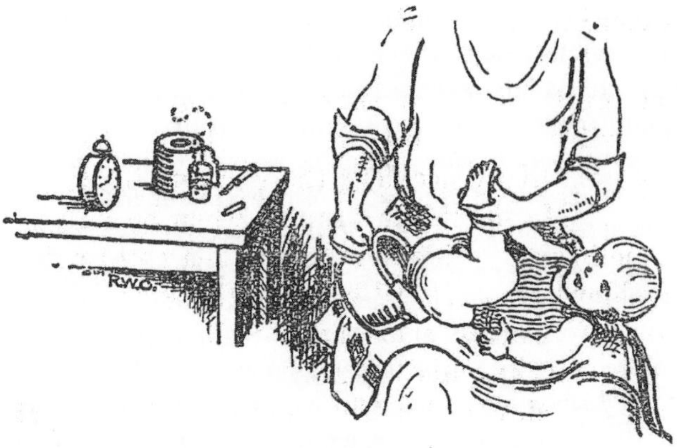

Start the baby with regular habits

Figure 6 Illustration depicting toilet training at end of first month of age. Children's Bureau, U.S. Department of Labor, *Infant Care*, 1935, 59

A child's reactions to eliminative functions will be largely conditioned by the attitudes of the adults in charge. Eliminative processes, involving as they do relief from bodily tensions, are normally pleasant and are almost always so regarded by the child. Yet many adults have themselves been conditioned to feel shame and disgust in connection with such functions; they must therefore re-condition themselves or else they will convey their own unfortunate attitudes to their children. (Blatz and Bott 137–8)

Anderson also emphasized the adoption of a "casual, matter-of-fact attitude" to facilitate the "development of control" over what he termed "a perfectly normal process, quite as natural as eating or sleeping" (60). Indeed, it would seem that a balancing act of freedom and restraint was called for, though always bound to fail, where proper bowel health was concerned. While it was important to "preserve the child's normal sense of satisfaction," one must also build up "inhibitions in regard to place and time of elimination" as well as prevent "unnecessary discussion in relation to the act" (Blatz and Bott 138). Moreover, advice abounded on how best to coordinate the disciplinary practices being acquired with the proper discourse – a discourse of strategic silences as well as carefully modulated utterances. Vigilance must be practiced about the linguistic surround that accompanied potty training; even before the child "begins to speak," we are told, the mother should "select the terms she wishes to use and employ them whenever she places the child on the toilet" so that the child himself will learn "through conditioning" to "employ them in anticipation of his needs" (Anderson 60).

Later, as the child became more deeply embedded in this discursive system, language needed to be deployed even more insistently to make him aware of the inner processes of his own body:

> The young child should be given some elementary explanation of the relation between intake of food and the elimination of waste, of the part played by food in the building up of the body, of the variety of ways in which waste products are eliminated, and of the importance for health of the unimpeded exercise of these functions. He will then understand urination and defecation as necessary parts of the bodily economy and therefore free from any suggestion of uncleanness or shame. Undoubtedly chronic constipation in adults may have certain psychological aspects. By training the child not only in right habits but in right *attitudes* we may help to avoid such later difficulties. (Blatz and Bott 138, emphasis in original).

As this passage shows, it was the role of the parent to facilitate the child's awareness not only of his own interiority, but especially of its economic nature, and of its place in the larger economy of production and consumption.

The establishment of 'habit clinics' in the late teens and early twenties coincided with this proliferation of child care literature. These clinics were perceived as institutional instances of, and safety nets for, the normal process of inculcating habit in the child (see Blatz and Bott 132). The 'home', as regarded from the point of view of the habit clinician, was "considered the workshop in which the personality of the child is being developed" (Thom 28). But when parents failed to control their negative affect, if they failed to maintain the proper discursive balance between restraint and satisfaction, then the household ceased to be a productive facility for the fabrication of the future citizen and resembled something more like a polluting miasma: "too frequently the home becomes the reservoir into which are poured all the resentment, disappointment, grief, and frustration associated with the unhappy experiences of the day." The children began to regard it as "the dumping ground of all unhappy and unhealthy adult emotions – a place used by all its inmates in their efforts to air the grievances against a world that has cheated them of something vital to their well-being" (Thom 28–9). It resembled, in other words, an emotional toilet rather than a 'workshop' where the child might develop the toilet habits that would ensure his or her own positive emotional and indeed moral development.

Just one paragraph from an article that outlined the history and role of the habit clinic in the United States demonstrates with a startling clarity how crucial regularity in early habit formation was considered to be:

> We begin to acquire habits at birth and continue to acquire them to a greater or less degree, depending upon our plasticity, until the end of life. The functions of eating, sleeping, and elimination become habitual. We develop habits of conduct toward those in authority. In some situations we acquiesce; in others we rebel. We have habits of conduct which include morals and manners. Our mental attitudes toward life are but habits of thought, and such traits as selfishness, shyness, cruelty, and fearfulness are merely emotional responses that have become habitual through repetition. Lying, stealing, and a disregard for the rights of others are likewise responses to certain situations, which through repetition become a part of the life pattern; in other words, they are habits. (Thom 26)

What becomes evident in this paragraph is that all aspects of human experience, from mundane bodily functions to intellectual, emotional, and moral activity, to traits that make up a personality were considered matters of 'habit'; moreover, it is clear from the literature on the training of habits in very young life that the regulation of the bodily functions, right down to the alimentary tract, was deemed crucial to the regulation of the individual as a social actor in the world. Thus, intense emphasis was placed on establishing good physiological habits, including those of elimination; to go astray in peristaltic regularity and efficiency was to risk delinquency of a more serious nature later on. "Bad habits such as enuresis and temper tantrums may offer great difficulties when it comes to treatment," one author wrote. But

> recent scientific researches in many fields have shown methods which are proving effective in solving these problems. These methods are being used by the Child Guidance Clinic to aid the child. They are especially effective with the normal child, helping him to develop free and unhampered by handicapping conflicts; to function at full efficiency in order that his debt to society may be adequately met by attainment of his maximum productivity (Van Norman Emery 663).

If he had trouble attaining an awareness of his own "bodily economy", and thus modulating the flow and ebb of that economy, the child could be brought by the responsible parent to the habit clinic, where he would receive the physiological (indeed, neurological) training that would help him to coordinate his own bodily economy with that of the state. Only thus could he "pay his debt" to society, in regularly issued instalments, as it were.

If the enteric system is a 'second brain', this brain acquires if not exactly a language of its own, then an intelligible, rationalized functionality in relation to the world outside it, and the world it is called on to process. The stimuli it takes in, assesses, and then acts upon are not incoherent, meaningless sights, sounds, smells, or sensations; rather, as is evident even in this short summary of toilet-training practices, the world with which the second brain comes into contact is regulated, deliberately structured to elicit correspondingly patterned responses, responses which are routed partly through the cognitive functions of the first brain, but largely also through the 'unconscious' (or rather, viscerally conscious) functions of the second brain. While I have been concerned with the disciplinary systems that characterized the early twentieth century, I hope that the reader will recognize in these pages a history of our own present, insofar as our peristaltic regularity, and the significance we attach to it, continues to be fashioned both inside the home and without. The project that lies before us is to undertake a thorough genealogy of this visceral training, as well as of the multiple and unexpected ways we manage, disrupt, or capitalize on it, as the peristaltic subjects of modernity.

Notes

1. Michael Gershon dates the birth of 'neurogastroenterology', a sub-field premised around the discovery that the "gut contains nerve cells that can ... operate the organ without instructions from the brain or spinal cord", to research conducted by Bayliss and Starling in the late 1890s, by Trendelenburg during the Great War, and by J.N. Langley who published *The Autonomic Nervous System* in 1921 (2–8). Meanwhile, following the motion studies of Étienne-Jules Marey and Eadweard Muybridge in the 1880s and 1890s, Frederick Taylor published *The Principles of Scientific Management* in 1914, while Frank and Lillian Gilbreth published their motion study books throughout the 1900s and 1910s. For a study of the gendered implications of the discourse around intestinal regularity and its relation to scientific management in the 1910s and 1920s, see my 'Female Peristalsis'. For a broader history of preoccupations with and anxiety about 'intestinal stasis' in this period, see Whorton, *Inner Hygiene*.

2. For an account of the emergence throughout the eighteenth and nineteenth century of the concept of the vegetative nervous system, and Langley's renaming of it as the autonomic nervous system in the early twentieth century, see Ackerknecht.

3. In his book, Gershon places emphasis on the gut's chemical processes rather than its potential for thinking and even learning; elsewhere, however, he seems to admit to the more speculative implications of a 'second brain'. Consider the anecdote he tells in a *New York Times* interview for instance:

> A big question remains. Can the gut's brain learn? Does it "think" for itself? Dr. Gershon tells a story about an old Army sergeant, a male nurse in charge of a group of paraplegics. With their lower spinal cords destroyed, the patients would get impacted.
>
> "The sergeant was anal compulsive." Dr. Gershon said. "At 10 A.M. everyday, the patients got enemas. Then the sergeant was rotated off the ward. His replacement decided to give enemas only after compactions occurred. But at 10 the next morning, everyone on the ward had a bowel movement at the same time, without enemas". Dr. Gershon said. Had the sergeant trained those colons? (Blakeslee para. 34–5)

4. See also Elizabeth Wilson, who argues "maybe ingestion and digestion aren't just metaphors for internalization; perhaps they are 'actual' mechanisms for relating to others. That is, perhaps gut pathology doesn't stand in for ideational disruption, but is another form of perturbed relations to others – a form that is enacted enterologically" (45).

Works cited

Ackerknecht, E.H. 'The history of the discovery of the vegetative (autonomic) nervous system.' *Medical History* 18.1 (1974): 1–8.

Anderson, John E. *Happy Childhood: The Development and Guidance of Children and Youth.* c.1933. New York: D. Appleton-Century, 1939.

Anderson, John E. and Marion L. Faegre. *Childcare and Training.* Minneapolis: Minnesota UP, 1929.

Blakeslee, Sandra. 'Complex and Hidden Brain in Gut Makes Bellyaches and Butterflies.' *The New York Times.* 23 January 1996. <http://query.nytimes.com/gst/fullpage.html?res=980CE0DF1F39F930A15752C0A960958260>

Blatz, William E. and Helen McMurchie Bott. *Parents and the Pre-school Child.* London: J.M. Dent, 1928.

Children's Bureau, U.S. Department of Labor. *Infant Care.* Bureau Publication No. 8. Washington: United States Government Printing Office, 1935.

Crum, Mason. 'Next Step – The Child Guidance Clinic.' *Hygeia: The Health Magazine* 13.10 (October 1935): 912–14.

Doane, Mary Ann. *The Emergence of Cinematic Time: Modernity, Contingency, the Archive.* Cambridge, Mass.: Harvard UP, 2002.

Foucault, Michel. *Discipline and Punish: The Birth of the Prison.* Trans. Alan Sheridan. NY: Random House, 1977.

Gershon, Michael. *The Second Brain: A Groundbreaking New Understanding of Nervous Disorders of the Stomach and Intestine*. New York: HarperCollins, 1998.

Massumi, Brian. *Parables for the Virtual: Movement, Affect, Sensation*. Durham: Duke UP, 2002.

Oldfield, Josiah. 'The Elimination of Constipation in Man.' *The Medical Press and Circular* 210 (15 December 1943): 381–5.

Thom, D.A. *Habit Clinics for Child Guidance*. Bureau Publication No. 135. Revised 1938 (first published in 1923 as *Habit Clinics for the Child of Preschool Age*). *Children's Bureau Studies*. Ed. William M. Schmidt. NY: Arno, 1974.

Van Norman Emery, E. 'The Child Guidance Clinic.' *The Pacific Coast Journal of Nursing* 21.11 (November 1925): 663–7.

Walton, Jean. 'Female Peristalsis.' *Differences: A Journal of Feminist Cultural Studies* 13.2 (Summer 2002): 57–89.

Whorton, James C. *Inner Hygiene: Constipation and the Pursuit of Health in Modern Society*. NY: Oxford UP, 2000.

Wilson, Elizabeth. *Psychosomatic: Feminism and the Neurological Body*. Durham: Duke UP, 2004.

13
Matter for Thought: The Psychon in Neurology, Psychology and American Culture, 1927–1943

Melissa M. Littlefield

> Invention, it must be humbly admitted, does not consist in creating out of void, but out of chaos; the materials must, in the first place, be afforded: it can give form to dark shapeless substances but cannot bring into being the substance itself
>
> Mary Shelley, Introduction to Third Edition
> of *Frankenstein*, 1831, 171

> Applied metaphysics assumes that reality is a matter of degree, and that phenomena that are more or less intensely real in the colloquial sense that they exist may become more or less intensely real, depending on how densely they are woven into scientific thought and practice
>
> Lorraine Daston, *Biographies of Scientific Objects*, 2001, 1

Scientific objects encourage us to imagine moments of discovery, collegial witnessing and citational webs. Yet, the process of creating narratives about scientific objects also enables and necessitates a certain kind of thinking: positing an origin that precludes previous instantiations (however different or varied); presuming that this origin has affects on future uses and developments of the term; ignoring the fragmentation, failures and accidents that surround and/or inform the creation of a concept; and pretending to provide an exhaustive history.

This article develops another kind of narrative, one that challenges attempts at linear, relational history by focusing on the rhizomatic emergence of a "novum" (Suvin 64): the psychon, psychome or *psychikon*. During the early decades of the twentieth century in America, the psychon emerged in neuroscience, psychology and science fiction (but not psychoanalysis) at multiple times and locations. William

Marston proposed his psychonic theory of consciousness in 1927, Stanley Weinbaum published a trilogy of science fiction stories concerning the psychonic theories of the fictional Dr. van Manderpootz between 1934 and 1935 and neuropsychiatrist Warren McCulloch and mathematician Walter Pitts detailed a theory of the psychon in 1943 that McCulloch claims to have conceptualized as early as 1923. None of the psychon's inventors referenced their contemporaries or predecessors, yet the psychon consistently fused morphology and ideology to create a unit of matter for psychology. In each variation, ending in -on or -ome, the term was aligned with such scientific concepts as chemistry's electron, physics's photon and – most significantly – neurology's neuron. The psychon described the least psychic event or unit of matter related to thought – but not fully encompassed by – nerves, synapses and reflex arcs.

Charting the constellation of psychonic invention during the early decades of the twentieth century need not entail a search for origins. Instead, it becomes an opportunity to "challenge the pursuit of origins" (Foucault 142). Moreover, the psychon's genealogy allows us to explore disciplinary and historical debates about the proper objects of several disciplines emerging and solidifying during the late modern era, including neurology and experimental psychology. As matter unit, the psychon was symptomatic of several tensions within modernity that illustrate a dynamic interaction of scale: the centrality of bodies and their components to the definitions of disciplinary boundaries, the fragmentation and augmentation of the body for social reform and the distillation and correlation of individual thoughts with a universal standard. Each of these tensions is imbued with ideologies of efficiency, the privileging of matter and anxieties about illusory appearances.

Of Novums and Materialist Battles

One way to read each emergence of an operative term like the psychon is through Darko Suvin's theory of the novum. Suvin's term is typically used to describe new language found in science fiction narratives. Yet, if we expand the scope and application of the 'novum', it is equally useful for discussing new terminologies introduced in science: "A novum of cognitive innovation is a totalizing phenomena or relationship deviating from the author's and implied reader's norm of reality," argued Suvin. He also noted that "the novum is a mediating category whose explicative potency springs from its rare bridging of literary and extraliterary, fictional and empirical, formal and ideological domains, in brief from its inalienable historicity" (64). The novum, then, bridges and connects the spheres of science and literature and draws on the larger cultural contexts from which science and literature conterminously emerge. Neither science nor fiction depends on the other for credibility, yet each owes a debt to "the scientific or cognitive horizon" within which they are "interpretable" (Suvin 65). I will detail a few of these scientific and cognitive contexts before returning more specifically to the psychon.

During the late nineteenth and early twentieth centuries, the human mind – its emotions, its links to the self and, most importantly, its internal mechanisms – was becoming the subject of several emergent sub-disciplines, including experimental psychology, psychiatry and psychoanalysis. In *The Mind of Modernism*, Mark Micale suggested that the development of various psychological schools between 1880 and 1940 was as momentous – and fractured – as the evolution of modernist art and literature. Although the modernist era is typically associated with psychoanalysis, Freud's theories were neither the most well-known, nor the most influential of the psychological schools. Indeed, "psychoanalysis was only one of many emerging models of mind that contributed to the constitution of the modern psychological self" (Micale 7). In the United States, for example, William James, G. Stanley Hall, E.B. Titchener and Hugo Munsterberg were equally influential figures who worked to establish the sub-discipline of experimental psychology.

Like the disciplines, historical eras and scientific theories represented here, the body is not a stable category; science, technology, literature and other cultural media consistently shape our understanding of what and how the body means.[1] Between 1890 and 1930, in both psychology and neurology, the body was undergoing a transition. On the one hand, the human was figured as a machine composed of functional units. This model was foundational to the creation of the nervous system and the construction of such atomized concepts as the neuron(e), germs, hormones and genes as independent units operating within the larger system of the body. On the other hand, English physiologists such as C.S. Sherrington introduced the complex systems model that focused on interactions among bodily systems and proposed that what Garland Allen termed 'dynamic materialism' should replace notions of mechanical materialism.[2] Studies of the body were split between delineating the body's fragments – its structure – and describing the complex interactions – the functional relationships – among its respective parts.

Experimental psychology's largest disciplinary development arose from the debates between Wilhelm Wundt and William James. Wundt, the preeminent German psychologist, was a proponent of structuralism, while James and his American cohort took a functionalist approach to psychology. The structuralists sought "general laws of behavior based on a coupling of sensory stimulation and self-report or introspection," while the functionalists were "interested in the way in which individuals adapted to their environment" (Landy 788). How, then should psychology pursue its future course?

During its formative years, experimental psychology was also haunted by its association with the so-called weaker disciplines, including philosophy.[3] How could psychology gain credence and acceptance as a distinct discipline? Should it ally itself with physiology and neurology or with the humanities? Wilhelm Wundt argued that psychology need not dialogue with the natural sciences, and was in fact, superior to them. In *Outlines of Psychology* (1897), Wundt explained why natural science was flawed: it "seeks to discover the nature of objects without reference to the subject. The knowledge that it produces is therefore *mediate* or *conceptual*. In place of the immediate objects of experience, it sets concepts gained from these objects by abstracting from the subjective components of our ideas" (5). In contrast, Wundt argued that psychology provided less symbolic and more direct access to the objects and subjects under its investigatory gaze. "Psychology...investigates the contents of experience in their complete and actual form, both the ideas that are referred to objects, and also the subjective processes which cluster about these ideas. The knowledge thus gained in psychology is, therefore, *immediate* and *perceptual*...all *concrete reality* is distinguished from all that is abstract and conceptual in thought" (5).

In sharp contrast, William James argued for psychology to be recognized as a natural science. In his 1892 essay, 'A Plea for Psychology as a Natural Science', he noted that at the very least, "we have here an immense opening upon which a stable phenomenal science must someday appear. We needn't pretend that we have the science already; but we can cheer those on who are working for its future, and clear metaphysical entanglements from their path. In short, we can aspire" (153). As James's language implies, psychology was not, unlike the natural sciences, seen as a credible science. A materialist angle was needed for psychology to 'aspire' to the supposedly more objective and quantitative knowledge of the physical sciences.

When experimental psychological laboratories began to be established in the 1880s, they were often modelled on those of the harder sciences. Studies of the mind, of interiority and the subjective interpretations of conscious states, were packaged in terms of bodily mechanics.

Psychophysicists, like Wilhelm Wundt, used quantitative instruments in their labs to measure perception and chart introspection as early as the 1880s. The American labs also focused on instrumentation, but they engaged less with introspective experiments on the mind. In fact, Hugo Munsterberg's description of the Harvard lab (founded by William James and directed by Munsterberg from 1892 until 1916) focused on detailing his instruments rather than his experiments to better authenticate the scientific work taking place in the lab. According to Munsterberg's unpublished papers, "a visit to a psychological institute would hardly suggest to the casual guest that it has anything to do with the mind" (qtd. Landy 788). After providing a brief description of the layout and equipment at Harvard, he went on to argue that "in short, everything suggests interest in bodily material processes, and nothing betrays the predominant activity of this scientific institute, the study of the mind" (qtd. Landy 789). Here, 'betraying' the true nature/purpose of the lab would apparently threaten the validity of experimental psychology. Indeed, as Munsterberg explained in 'The Psychological Laboratory at Harvard University' (1893), experimental psychology's purpose needs to be explained to external audiences because it "is too often confused with experiments upon the brain by vivisection, with hypnotism, and even with spiritism [sic]" (4).

In creating its materialist image, experimental psychology was responding to the status accorded anatomy, physiology and another solidifying science: neurology, a discipline that "employs causal-mechanical or mechanistic explanations" (Revonsuo 48) at least in this period.[4] During the late nineteenth century, while Munsterberg worked at Harvard, the concept of the nervous system came into being and the neuron emerged as a disciplinary unit of matter in 1891[5] thanks to the work of Heinrich Wilhelm Gottfried Waldeyer-Hartz. The response of several psychologists was to create a competing unit of matter: the psychon.

Scholars have argued that the invention of the psychon could be linked to Freud's 'Project for a Scientific Psychology' (1895).[6] I would argue that although Freud predicted the trend towards material explanations of mind and responded with his 'Project' in 1895, he relied on neurology's unit of matter, the neuron. Initially, Freud titled his project 'The Psychology for Neurologists' (*Complete Letters* 127), explicitly linking his concepts to the burgeoning, qualitative discipline. Later, in a letter to Wilhelm Fliess, Freud explained that one of his "intentions" was "to discover what form the theory of psychical functioning will take if a quantitative line of approach, a kind of economics of nervous force, is introduced into it" (*Complete Letters* 129). His aspirations materialized in very specific relation to the natural sciences as he wrote his 'Project'. In the opening paragraphs of this document, which he later sacked, Freud argued that his "intention is to furnish a psychology that shall be a natural science: that is, to represent psychical processes as quantitatively determinate states of specifiable material particles, thus making those processes perspicuous and free from contradiction.... The neurons are to be taken as the material particle" ('Project' 295). Thus, Freud's 'Project' is "ostensibly a neurological document" (Strachey 290). Freud's eventual abandonment of the neuron and his larger 'Project' in favour of psychoanalysis was a consequence of the common conclusion that neurological substrates could not, ultimately, account for consciousness. While Freud moved away from matter units as he created psychoanalysis, other psychologists, neurologists, philosophers and science fiction authors invented the psychon, a matter unit for psychology that could legitimate psychology as a natural science and account for consciousness.

Body wars: The psychon as contested unit

The psychon, as term and concept, emerged because it seemed so plausible, ideologically and morphologically. It mirrored other atomist concepts, while also bridging and linking conceptions of the imagined and the real, the immaterial and the material that informed the early decades of the twentieth century. An examination of the psychon reminds us that constructions of bodily boundaries are mutually imbricated with disciplinary boundaries. For psychology, the division between the structural and the functional, the material and the immaterial, was often expressed in terms of disciplinary turf wars. It is not coincidental that the psychon was consistently linked to the disciplinary status of psychology during the decades following the development of the nervous system, the emergence of the neuron and the split between the American Philosophical Association and what would become the American Psychological Association in 1892. During these disciplinary developments, the psychon became an attempt to legitimate psychology while also strengthening and policing its disciplinary boundaries.

Auguste Forel made such a move towards establishing the discipline by introducing a new unit of matter, the 'psychome', in *Hypnotism; Or Suggestion and Psychotherapy* (1906).[7] The entity was intended "to express each and every psychical unit" (6). Yet, the term was immediately discounted by the author who appended his reservations in a footnote: "The author apologizes for this term. He has introduced it for brevity's sake" (6n1).[8] Forel's hesitations are notable because the psychome is not a morphological oddity; indeed, its structure, function and fragmented nature are consistent with the atomism of the era. Arguably, Forel's discomfort has more to do with the psychome's immateriality. Unlike the neuron(e), another of Forel's inventions, the psychome is not a material object; it is a hypothetical term – a conceptual placeholder – introduced for 'brevity's sake' from which Forel spun his larger theories about mind, brain and matter. Indeed, Forel's chagrin signifies the marginality of the psychome: Is it a concept or an object? Does its use validate or denigrate scientific inquiry? Is it a concept waiting to be legitimated by material and visual confirmation or a term of convenience without substantive meaning or appeal?

Despite Forel's censure, the psychome did not disappear into the footnotes of history. Over the next century, a related entity, the psychon, emerged and continued to circulate in neuroscientific, psychological and literary discourses. In addition to Stanley Weinbaum's and Warren McCulloch's versions, the psychon can also be found in psychology texts by Henry Lane Eno (1920) and Paul and W.R. Bousefield (1927). But it was one of Hugo Munsterberg's students, William Marston,[9] who used the psychon (in his role as the most outspoken defender of psychology's new disciplinary boundaries and its legitimacy) as a distinct approach to understanding the mind/brain. Marston's initial use of the term occurred in *The Emotions of Normal People* (1928) and was subsequently revived in his psychology textbook *Integrative Psychology: A Study of Unit Response* (1931). He introduced the psychon as both a unit of matter and a conceptual entity through which consciousness could be described and measured. In so doing, Marston reiterated Forel's incarnation without the latter's sense of mortification. Indeed, Marston stalwartly defended his creation as the legitimate structural unit of psychology.

Marston's conceptualization of the psychon presumed several postulates: 1) psychology's proper object of study is consciousness; 2) "consciousness is to be identified with synaptic energy" (46), specifically with "the totality of energy generated within the junctional tissue between any two neurons" (52); and 3) the psychon described "any particular unit of junctional tissue which may be under discussion" (52). In short, the psychon was a way to describe the energy transferred between neurons – energy that, in its ultimate translation, is equal to consciousness itself.

Marston's definition of an immaterial concept that has material effects resonated with the two other appearances of the psychon during the 1920s and stood in contradistinction to an earlier form of the term. In *Activism* (1920), Henry Lane Eno argued for the existence of the psychon based on the logic of atomism. His elaborate description moved from the known to the speculative. Beginning with atoms, which are composed, in part, of electrons, Eno then moved to "still more fundamental units ... of a higher plane, whose characteristic activity stands in a relation to electricity analogous to that in which electricity stands to matter" (45). Eno defined these "hypothetical units ... of higher activity in the psychic processes" as psychons (45).

Paul and W.R. Bousfield's *The Mind and Its Mechanism* (1927) could arguably be read as an extension of Eno's text. It presumed a similar ordering of value: the physical world of electrons and protons, and an alternate, perhaps more fundamental unit that could be derived from these known quantities. "Ethereal protons and electrons we know as the constituents of matter.... As the basis of the 'immaterial substance' we may postulate a second order of 'ons' which are, like protons and electrons, fashioned out of the ether. Let us call these 'ons' by the name of 'psychons'" (22). In the same paragraph, the Bousfields addressed questions about the psychon's physicality more specifically than Eno by drawing on established knowledge – and, more importantly, established ignorance – about a first order of 'ons': electrons and protons. The rhetorical question that begins this segment of the argument ("What do we mean by 'immaterial'?") precedes a description that emphasized the era's atomism:

> Simply that the mass of the psychon is of the second order of small quantities compared with the mass of an electron – so small as to be inappreciable by the methods by which we appreciate the mass of the material atom.... We conceive therefore of psychons as being in the physical realm, though immaterial, but they may also be regarded as being in a borderland which connects the physical and the psychic. (22–3)

With this description, we return again to the psychon as a hypothetical concept: one that could, as it is figured here, help negotiate the role of materiality. Theirs was a definition which at least recognizes the precarious position of psychology in an era devoted to the prevalence of natural science and the mechanization of the body. Yet, neither Eno nor the Bousefields directly addressed psychology's rival discipline of neurology. Other scientists, such as William Marston, made the explicit and crucial comparison.

In addition to an immaterial entity with material ramifications, Marston's psychon was a direct response to those who would denounce psychology as a fatally flawed discipline doomed to obscurity given the rise of material science. First, Marston specifically differentiated and valued his psychon as distinct, rather than relational: "Propagation of energy upon any psychon, or unit of junctional tissue, is *definitely dissimilar in nature* to the passage of nervous energy through individual neurons" (Marston 52, emphasis added). Elsewhere, as he responded conversationally to charges that psychology is a dying discipline, Marston used this notion of difference as a powerful rhetorical strategy. "If, then, there exists no further type of matter-unit capable of modifying neurotic behavior, psychology, for all I can see, is out of a job," Marston goaded. "Should I become convinced of this state of facts I should feel compelled to consider psychologists in the same relation to neurologists as are carpenters to architects, and I should, for my own part, try to escape the fixed limits of craftsmanship by studying my way into the ranks of my immediate intellectual superiors" (47). The biting sarcasm of this passage is remarkable in part because of what was at stake for psychology and neurology, including their relative complexity, relation to one another, scope, and proper objects of knowledge. Marston more explicitly linked neurology to the study of the built environment, calling its object the "tungsten filament in an electric light bulb," while simultaneously figuring psychology's object as the more complex process of "illumination" and the "generation of consciousness" (52).

Psychology needed the psychon, but so did neurology, if scientists were to ultimately understand the workings of the mind/brain. After revelling in cynicism, Marston quickly returned to his pointed argument about the psychon's existence and value:

> if there exists still another sort of matter unit beyond the neurone, capable of undergoing its own particular series of changes called 'conscious' or 'psychical' changes, and capable of modifying, by these changes, the behavior of neurons, then, and then only is psychology truly justified in assuming a definite place among the physical sciences by the side of physiology and neurology. (47)[10]

Here, Marston not only justified psychology's need for the psychon, but also linked its existence to our understanding of neuronal processing.

The psychon was not merely hypothetical; instead, it affected the material structures claimed by the purview of neurology. Marston reversed the hierarchy of psychology and neurology, psychical and material sciences and the immaterial and the tangible. His claim constructed a crucial role for psychology in the study of the mind, revalued knowledge claims produced by neurology and reassigned the triggers of brain/mind processes to psychology instead.

Like Marston, Warren McCulloch's work represents a hybrid of several disciplines: neurophysiology, neuropsychiatry, psychology and philosophy. He is best known for his work with mathematician Walter Pitts: 'A Logical Calculus of the Ideas Immanent in Nervous Activity' (1943), a document that helped generate the debates about Artificial Intelligence, the Macy Conferences on cybernetics, and debates about the functioning of animals, humans and machines. In this paper, McCulloch and Pitts characterized neuron activity as all-or-none, an idealized way to speak about neurons as 'on or off'. McCulloch's psychon served to connect these binarized isomorphic units, explaining how neurons communicated with each other.

Simultaneously, the psychon was drafted to create relations and divisions between psychology and neurology. McCulloch explained that his theory of logical networks could be of use to psychology and neurology; to do so, he linked his neural networks to the "simplest psychic act; you may call it a 'psychon' if you will" (392). Note that McCulloch was also sceptical about conjuring the psychon even as he described the networked connections between neurons – connections which are themselves immaterial. In the same breath that McCulloch invoked the psychon, he discredited it as a concept whose materiality belonged to the neuron: "even if the analysis was pushed to the ultimate psychic units or 'psychons,'...a psychon can be no less than the activity of a single neuron. Since that activity is inherently propositional, all psychic events have an intentional, or 'semiotic,' character" (37). Here, the psychon was no longer a motivating force, as it was for Marston; instead, the psychon's and neuron's functions were elided. The neuron remained material; the psychon was relegated to semiotic status: a signifier not directly connected to any signified.

McCulloch's narrative about the psychon, like most stories, has a history. Although he published 'A Logical Calculus' in 1943, McCulloch claimed to have invented the concept of the psychon as early as 1923, placing him squarely in the midst of the psychon's busiest decade alongside Paul Eno, P. and W.R. Bousefield and William Marston. "In 1923," wrote McCulloch, "I gave up the attempt to write a logic of transitive verbs and began to see what I could do with the logic of propositions. My object, as a psychologist, was to invent a kind of least psychic [to mean of the psyche] event, or 'psychon' " (8).[11]

He went on to describe the psychon in terms of atomism (highlighting fragmentation and structure) and connectionism (highlighting functionality, complexity and association). McCulloch's psychon "was to be to psychology what an atom was to chemistry or a gene to genetics" (McCulloch 392). He also insisted that "my psychon differed from an atom and from a gene in that it was not to be an enduring, unsplittable object, but a least psychic event.... My postulated psychons were to be related much as offspring are to their parents, and their occurrence was in some sense to imply a previous generation that begat them" (392). McCulloch's recollection of his 1923 psychon revealed his desire "to make psychology into experimental epistemology" by exploring what he called the "embodiment of mind" (389). First, he, like Marston, Eno and the Bousefields, insisted that psychology needed a unit of matter, an '-on' if it is to establish compelling disciplinary boundaries based on a proper and material object of knowledge. However, McCulloch constructed the psychon as immaterial, ephemeral, fleeting: an event or function rather than a substance. This second hypothesis, thus, avoided any discussion of an origin, material or otherwise, speaking instead in terms of activity, predecessors, and proposed futures. McCulloch's hypothesis "gave a theory of activity progressing from sensation to action through the brain.... The implications of psychons pointed to the past, and their intention foreshadowed the proposed response" (McCulloch 392). McCulloch's embodiment of mind invoked transitory events, action and function as opposed to material objects, and was symptomatic of the phase shift between structure and function, body and mind taking place in science and culture more broadly, as I will go on to discuss.

What's the matter: Functional fragments

As we have seen, the psychon was both a "fictitious neurological substrate" (Burnham 31) and a functional event that affected the structural units of the nervous system. It was both a product of the modern impulse to segment, calibrate and correct the body and a challenge to the mechanistic impulse that produced Taylorism, Fletcherism and Fordism. In this section, I want to explore the psychon's integration into debates over structure and function, body and mind as it was deployed for social reform through bodily management.

Systems of bodily management are based on an idealized conception of the body, one that does not suffer from the crises associated with modernization and modern civilization such as the threat of irrationality and fatigue.[12] This ideal body was perfectly rational, normalizable, able to withstand fatigue and was not excessive (physically or emotionally). However, when real bodies could not or did not conform to such ideals, modern techniques of bodily management intervened, though they did so via systems of corporeal and psychical fragmentation. Scientists across a wide range of disciplines were invested in reducing systems to their component parts. This pursuit, fuelled by the idea that "knowledge of components led to knowledge of causes" (Burnham 12), engendered the discovery of cells, germs, hormones and genes. The newly constituted individual was further splintered into various objective, quantifiable and controllable components that could be monitored and modified by specifically developed technologies. Both Marston and McCulloch were invested in the psychon as just such a disciplinary object, but also as a liminal entity. They were also ultimately interested in social intervention. In the wake of behaviourists and the James-Lang theory of emotions, the body took centre stage, relegating the mind to mere reflex influenced by environment, action and conduct. As a unit of matter, the psychon was product and producer of a controversial hypothesis: that the mind could affect the body. If modernity exhibits the "desire to intervene in the body; to render it part of modernity by techniques which may be biological, mechanical or behavioral" (Armstrong 3), the postulation of the psychon provided hope for troubled minds and endowed psychology and neurology with the means to intervene. Under this rubric, diagnoses could be made and treatments offered.

For McCulloch, reform was tied more directly to the mental hygiene movement.[13] During these decades, psychology and psychiatry were dominated by two movements: mental testing (IQ) and mental hygiene (post-First World War psychotherapy). During the height of the mental hygiene movement, between and following the World Wars, science had high hopes for treating non-institutionalized, but nonetheless pathological populations, including returning soldiers. The movement defined disease as "a product of environmental, hereditarian, and individual deficiencies; its eradication required the fusion of scientific knowledge and administrative action" (Grob 144). To combat the evils of modern life that caused mental imbalance, proponents of the mental hygiene movement initiated campaigns for sterilization and against venereal diseases, instated educational programs for children and intervened in behavioural problems such as alcoholism.

In 'A Logical Calculus', McCulloch argued that because of the psychon and his larger theory of networks, "diseased mentality can be understood without loss of scope or rigor, in the scientific terms of neurophysiology. For neurology, the theory... clarifies the relations of disturbed structure to disturbed function" (38). As a point of connection between disciplines, the psychon allowed for reform on multiple fronts simultaneously. For example, McCulloch also noted that the psychon ensures that

> both the formal and the final aspects of that activity which we are wont to call *mental* are rigorously deducible from present neurophysiology. The psychiatrist may take comfort from the obvious conclusion concerning causality – that, for prognosis, history is never necessary... for the psychiatrist it is more to the point that in such systems 'Mind' no longer 'goes more ghostly than a ghost.' (38)

By demystifying the mind, McCulloch was working within the paradigm of mental hygiene movement and against more traditional notions of prevention. Instead of arguing that psychiatrists could only deduce pathology by examining the causal relationship between behaviours and disease, McCulloch urged his readers to see disease as a material phenomenon that could be best identified and prevented by the intervention of science and the proper administrative initiatives.

For Marston, reform involved emotional re-education and a revaluation of 'normal' emotions. His proposals were based on principles that informed the concept of the psychon: that mind, mental processes and consciousness affect the body can be measured in the autonomic nervous system and ultimately adjusted. One of the main reasons Marston was interested in systemic change was because larger institutions, traditions and cultural mores had separated individuals from their so-called normalcy. Thus, we must begin uncovering and reclaiming our secret, but normal lives through "emotional re-education" (Marston 391). For Marston, this meant accepting a psychological norm uninfluenced by social institutions and cultural mores: "the only practical reeducation consists in teaching people that there is a norm of psycho-neural behavior, not dependent in any way on what their neighbors are doing or upon what they think their neighbors want them to do. People must be taught to love parts of themselves, which they have come to regard as abnormal" (391).

Marston's and McCulloch's conceptions of social reform were both relational: for both there are connections between structure and function and between the individual and society. The psychon encouraged not only these kinds of connections, but also served to identify anxieties about making thoughts matter. In my final section, I explore how science fiction's psychon more aptly revealed a set of tensions unexplored in the scientific literature of the era.

Making thoughts matter: Visualization, atomism and literature

As a novum, the psychon crossed both scientific disciplinary boundaries and those erected between the humanities and the sciences. Because literature identifies and embodies similar inquiries emerging from modern, materialist, atomist culture, several stories, including Paul Ernst's 'From the Wells of the Brain' and Stanley Weinbaum's van Manderpootz trilogy, directly addressed the materialization of thought. Weinbaum's stories are of particular interest because in them the psychon played a staring conceptual role, without reference to any psychologist or neurologist of the era. Through other scientific citations, including Einstein's work on relativity, Weinbaum maintains that his literature was completely connected to the world of science; but it was his creation of the psychon as novum that served to connect the era's literature, neurology and psychology. These stories literally represented ideas about materialism and atomism – thought as matter – which revealed anxieties and questions about the nature of thought and our ability to distil and correlate individual thoughts with a universal standard.

Stanley Weinbaum's version of the psychon emerged in 1935 during his brief stint as a pulp science fiction writer. In 'The Ideal', the second instalment of his Professor van Manderpootz trilogy, Weinbaum introduced a novum with a familiar morphological structure: the psychon. According to van Manderpootz, a stereotypically minutia-obsessed scientist, the psychon "is one electron plus one proton, which are bound so as to form one neutron, embedded in one cosmon, occupying a volume of one spation, driven by one quantum for a period of one chronon. Very obvious; very simple" (225). While van Manderpootz's description is far from "[v]ery obvious" or "very simple," his description *is* an obvious parody of the era's atomism: that desire to fragment the body into component parts that can be more readily studied.

Dr. van Manderpootz's psychon enables him literally to convert thought into matter, which can then be represented visually on a screen. The process works through the 'idealizer': a technological contraption that extracts psychons from any given participant and converts them into other matter units like "quanta" or "cosmons." This process is directly compared to the ways "a Crookes tube or X-ray tube transforms matter to electrons" (225). In this respect, Weinbaum connects his fictional doctor's science to the popular theories of the era.[14] The visual representations produced by van Manderpootz's 'idealizer', though, are not simply thoughts on screen; they are supposed to be ideal Aristotelian-like 'forms' that are not specific to the individual. In an extended explanation, the doctor explains the process in these terms:

> I will make your thoughts visible! And not your thoughts as they are in that numb brain of yours, but in *ideal* form. Do you see? The psychons of your mind are the same as those from any other mind, just as electrons are identical, whether from gold or iron.... Therefore, my idealizer shows your thoughts released from the impress of your personality. It shows it – ideal! (225)

The ideal van Manderpootz speaks of makes good sense, given the atomism and materialism of the era: matter is matter is matter no matter whose brain produces it. The materialization of thought enables both the fragmentation and augmentation of the body we saw in Marston and McCulloch's work.

In Weinbaum's scheme, however, the atomization and utilization of bodily fragments took on a slightly different tenor. Doctor van Manderpootz's distillation of individual thoughts does not produce ideal forms, but unique ideals informed by past experiences and, more importantly, the very process of extracting thought. When Dixon, the playboy narrator of 'The Ideal', agrees to participate in van Manderpootz's experiment, he envisions not *the* ideal woman, but *his* ideal woman. This figure, we are told, is a combination of several things: women he has loved, visions of the feminine he was exposed to as a child and, eventually, previous instantiations of his envisioned ideal. "The very fact that I had seen an ideal once before had altered my ideal, raised it to a higher level," Dixon notes, "[w]ith that face among my memories, my concept of perfection was different than it had been" (231). Even and especially when thoughts are materialized, they are not, and do not remain, static, universal or repeatable. Instead, they are affected by the very processes of distillation.

Instead of offering unmitigated access and standardized results, 'The Ideal' asked readers to review the psychon's appearance in the early twentieth century against a set of new questions and lines of inquiry: the belief that component parts could explain the whole, that materialism promises to bring inquiry down to a level in which nothing can be hidden and that a universal unit of matter could provide the means to understand and codify individual thoughts and states of consciousness. These ideals of transparency promised certain knowledge in the place of opacity, unarticulated motivations and individual deviance. However, in experimental and imagined practice, individuals are unknowns, fallible and imperfect. The psychon was a unit of matter that could explain a principle, serve as a proper disciplinary object, and provide fodder for several cultural ideals; but its emergence in neurology, psychology and science fiction revealed more about larger cultural anxieties than durable units of matter.

Conclusion

As novum, invention, concept, term and unit of matter, the psychon emerged out of the fabric of an era obsessed with atomism and materialism. Its history is multiple, changeable, archival and contemporary; but each instantiation of the psychon reminds us that this unit of matter was and continues to be fashioned out of several common ideologies: that the fragmentation of the body leads to certain knowledge about and possibilities for augmentation; that bodily and disciplinary boundaries are mutually imbricated; and that universal principles and standards could create transparency.

A linear narrative about the psychon would ignore the fraught, tenuous and multiple instantiations of the concept that were forgotten and/or discarded over the course of several decades. A genealogy of the psychon as novum preserves a sense of contingency and asks us to consider how objects come into and out of being. This genealogy, in particular, reminds us that coincidences are often better described as constellations: clusters of activity linked to larger cultural concerns of an era. In this case, neurologists, psychologists, philosophers and science fiction writers independently and collaboratively invented a term and concept that dovetailed with modern anxieties about disciplinary status, social reform and the transparency of individual thought.

Most importantly, the psychon illustrates that theorizations of the mind/brain/body extended far beyond the popular psychoanalytic visions of the modern era. As a concept, the psychon crossed multiple disciplinary boundaries, engaged consciousness as material phenomenon and exemplified the multiplicity of approaches to the mind's structure and function. Did the raw materials of modernity produce psychoanalysis? Yes. But this chaos also engendered experimental psychology and neurology, concepts like the psychon and larger social reform movements in science and modern society.

Notes

1. See Donna Haraway's *How Like A Leaf; Modes Witness* and *Simians, Cyborgs and Women*, Emily Martin's 'The Egg and the Sperm' and *The Woman in the Body*, Nelly Oudshorn's *Beyond the Neutral Body* and Tim Armstrong's *Modernism, Technology and the Body*. Culture plays a role here, but I don't want to segregate culture from science and technology – they are interrelated concepts.
2. For a complete explanation of the shifts that took place between the end of the nineteenth and beginning of the early twentieth century, see John Burnham's *Paths into American Culture*, particularly 'The Mind–Body Problem in the Early Twentieth Century.'
3. Concerning the term 'new psychology', see Steven Ward's *Modernizing the Mind*.
4. For more information on the history of neurology more generally, see Clarke and Jacyna, and Machamer et al., eds, particularly Olaf Breidbach's article in the collection.
5. It should be noted that the term 'nerve cell' had been used since the 1830s, but the term 'neuron' was coined in 1891 on the basis of the recent attainment of microscopic evidence that nerves were, in fact, cells.
6. Mark Germine alludes to connections between Freud and the psychon based on the abstraction entailed by various types of matter units: "Freud identified the neuron as the fundamental model of all mental activity, describing the function of the neuron principally on the basis of the spinal reflexes. This type of abstraction, embodied later in the concept of the 'psychon,' was to become a founding principle of connectionist models of the mind" (81). James Strachey extrapolates further, noting in his introduction to the 'Project' that "it has been plausibly pointed out that in the complexities of the 'neuronal' events described here by Freud, and the principles governing them, we may see more than a hint or two at the hypotheses of information theory and cybernatics in their application to the nervous system" (292).
7. This was not the first appearance of this term; but the first modern appearance. For ancient references, see Galen's *pneuma psychikon* (Rousseau, *Nervous Acts*).
8. August Henri Forel (1848–1931) was a Swiss entomologist, neuroanatomist and psychiatrist. In collaboration with Wilhelm His and Fridtjof Nansen, he introduced the modern neuron theory in 1887. He, like Marston and McCulloch also worked on social hygiene projects, including the transmission of sexually transmitted diseases and alcoholism.

9. William Marston invented the lie detector test during his graduate work at Harvard between 1917 and 1921. He had a long and varied career that also included the creation of Wonder Woman in 1941.

10. Given Freud's rejection of matter units, Marston's insistence on a unit of matter for psychology is an important move that, again, differentiates the psychon and its inventors from psychoanalysis and arguments for any singular disciplinary trajectory.

11. McCulloch's complete definition included the following "properties" of the psychon: "First, it was to be so simple an event that it either happened or else did not happen. Second, it was to happen only if its bound cause had happened... that is, it was to imply its temporal antecedent. Third, it was to propose this to subsequent psychons. Fourth, these were to be compounded to produce the equivalents of more complicated propositions concerning their antecedents" (8; qtd. in Abraham 7).

12. See Tim Armstrong's *Modernism, Technology and the Body*, Nancy Cartwright's *Screening the Body* and Anson Rabinbach's *The Human Motor* for discussions of questions of modernity, bodily fragmentation, augmentation, visualization and management.

13. For more information on the history of the mental hygiene movement, see Grob, especially Chapter 6. For affects and focus on children, see Richardson. Important primary, historical documents include Clifford Beers' *The Mental Hygiene Movement* (1923), *Handbook of the Mental Hygiene Movement and Exhibit*, prepared by the National Committee for Mental Hygiene (1913), and William White's 'The Origin, Growth and Significance of the Mental Hygiene Movement' (1930).

14. I am indebted to the editors of this volume for pointing out that the visualization of thought in science fiction has an even longer history than this chapter accounts for. For an earlier fictional fantasy of thought represented visually by technological means, see Pollock.

Works cited

Abraham, Tara. '(Physio)logical Circuits: The Intellectual Origins of the McCulloch-Pitts Neural Networks.' *Journal of the History of the Behavioral Sciences* 38 (2002): 3–25.

Allen, Garland. *Life Science in the Twentieth Century*. Cambridge: Cambridge UP, 1978.

Armstrong, Tim. *Modernism, Technology and the Body: A Cultural Study*. Cambridge: Cambridge UP, 1998.

Beers, Clifford. *The Mental Hygiene Movement*. New York: Doubleday, Page & Co., 1923.

Bousefield, P. and W.R. Bousefield. *The Mind and its Mechanism*. New York, E.P. Dutton, 1927.

Burnham, John. *Paths into American Culture: Psychology, Medicine, and Morals*. Philadelphia: Temple UP, 1987.

Cartwright, Lisa. *Screening the Body: Tracing Medicine's Visual Culture*. Minneapolis: Minnesota University Press, 1995.

Clarke, Edwin, and L. S. Jacyna. *Nineteenth-Century Origins of Neuroscientific Concepts*. Berkeley: California UP, 1987.

Daston, Lorraine, ed. *Biographies of Scientific Objects*. Chicago UP, 2000.

Eccles, J. C. 'Do Mental Events Cause Neural Events Analogously to the Probability Fields of Quantum Mechanics?' *Proceedings of the Royal Society of London. Series B, Biological Sciences*, 227.1249 (22 May 1986): 411–28.

——. 'Evolution of Consciousness.' *Proceedings of the National Academy of Sciences* 89 (1992): 7320–4.

Eno, Henry Lane. *Activism*. Princeton: Princeton UP, 1920.

Ernst, Paul. 'From the Wells of the Brain.' *Astounding Stories* 35 (October 1933): 48–54.

Forel, Auguste. *Hypnotism; or, Suggestion and Psychotherapy; A Study of the Psychological, Psycho-physiological and Therapeutic Aspects of Hypnotism*. Trans. H. W. Armit. New York: Allied, 1949.

Foucault, Michel. 'Nietzsche, Genealogy and History.' *Language, Counter-Memory, Practice: Selected Essays and Interviews*. Ed. D. F. Bouchard. Ithaca: Cornell UP, 1977.

Freud, Sigmund. 'Project for a Scientific Psychology.' 1895. *The Standard Edition of the Complete Works of Sigmund Freud, Vol. 1*. Trans. James Strachey and Anna Freud. London: Hogarth Press & Institute of Psycho-Analysis, 1966. 283–397.

——. *The Complete Letters of Sigmund Freud to Wilhelm Fleiss, 1877–1904*. Trans. Jeffery Moussaieff Masson. Cambridge, MA: Belknap Press of Harvard University, 1985.

Germine, Mark. 'The Concept of Energy in Freud's *Project for a Scientific Psychology*.' *Annals of the New York Academy of Science* 843 (1998): 80–90.

Grob, Gerald N. *Mental Illness and American Society, 1875–1940*. Princeton: Princeton UP, 1983.

Haraway, Donna. *How Like a Leaf: An Interview with Thyrza Nichols Goodeve / Donna J. Haraway*. New York: Routledge, 1999.

——. *Modest Witness@Second-Millennium.FemaleMan-Meets-OncoMouse: Feminism and Technoscience*. New York: Routledge, 1997.

——. *Simians, Cyborgs and Women: The Reinvention of Nature*. New York: Routledge, 1991.

James, William. 'A Plea for Psychology as a Natural Science.' *Philosophical Review* 1 (1892): 146–53.

Karczman, Alexander. 'Sir John Eccles, 1903–1997; Part 2. The Brain as a Machine or as a Site of Free Will?' *Perspectives in Biology and Medicine* 44.2 (2001): 250–62.

Landy, Frank. 'Hugo Munsterberg: Victim or Visionary.' *Journal of Applied Psychology* 77 (1992): 787–802.

Machamer, Peter, Rick Grush and Peter McLaughlin, eds. *Theory and Method in the Neurosciences*. Pittsburgh UP, 2001.

Marston, William. *Emotions of Normal People*. New York: Harcourt, Brace, 1928.

Marston, William, C.D. King, and E.H. Marston. *Integrative Psychology; A Study of Unit Response*. London: Harcourt, Brace, 1931.

Martin, Emily. 'The Egg and the Sperm: How Science Has Constructed a Romance Based on Stereotypical Male–Female Roles.' *Signs* 16.3 (1991): 485–501.

——. *The Woman in the Body: A Cultural Analysis of Reproduction*. Boston: Beacon, 1992.

McCulloch, Warren. *Embodiments of Mind*. Cambridge, MA: MIT Press, 1988.

Micale, Mark. *The Mind of Modernism: Medicine, Psychology, and the Cultural Arts in Europe and America, 1880–1940*. Stanford: Stanford UP, 2004.

Munsterberg, Hugo. *The Psychological Laboratory at Harvard University*. Cambridge: Cambridge UP, 1893.

National Committee for Mental Hygiene. *Handbook of the Mental Hygiene Movement and Exhibit*. New York: The National Committee for Mental Hygiene, 1913.

Oudshorn, Nelly. *Beyond the Natural Body: An Archeology of Sex Hormones*. New York: Routledge, 1994.

Pollock, Walter Herries. 'The Phantasmatograph.' *Longman's Magazine* 34.199 (May 1899): 58–69.

Rabinbach, Anson. *The Human Motor: Energy, Fatigue and the Rise of Modernity*. New York: Basic Books, 1990.

Revonsuo, A. 'On the Nature of Explanation in the Neurosciences.' *Theory and Method in the Neurosciences*. Eds. P. K. Machamer, R. Grush and P. McLaughlin. Pittsburgh: Pittsburgh UP, 2001. 45–69.

Richardson, Theresa R. *The Century of the Child: The Mental Hygiene Movement and Social Policy in the United States and Canada*. Albany, NY: State University of New York Press, 1989.

Rousseau, George. *Nervous Acts: Essays on Literature, Culture and Sensibility* New York: Palgrave, 2004.

Shelley, Mary. 'Introduction to Third Edition (1831).' *Frankenstein*. New York: W. W. Norton, 1996.

Strachey, James. 'Editor's Introduction to Freud's *Project for a Scientific Psychology*.' *The Standard Edition of the Complete Works of Sigmund Freud, Vol. 1*. Trans. James Strachey and Anna Freud. London: Hogarth Press & Institute of Psycho-Analysis, 1966. 283–93.

Suvin, Darko. *Metamorphoses of Science Fiction: On the Poetics and History of a Literary Genre*. New Haven: Yale UP, 1979.

Ward, Steven. *Modernizing the Mind: Psychological Knowledge and the Remaking of Society*. Westport: Praeger Publishers, 2002.

Weinbaum, Stanley. *A Martian Odyssey and Other Science Fiction Tales: The Collected Short Stories of Stanley G. Weinbaum*. Westport: Hyperion, 1974.

White, William. 'The Origin, Growth and Significance of the Mental Hygiene Movement.' *Science* 17.1856 (1930): 77–81.

Wundt, Wilhelm. *Outlines of Psychology*. Leipzig, 1887.

Index